THE BRITISH BIG CAT PHENOMENON

SIGHTINGS, FIELD SIGNS, AND BONES

JONATHAN MCGOWAN

CONTENTS

SYNOPSIS

Born in Bournemouth during the nineteen sixties and brought up in many areas of Dorset, I have gotten to know my county like the back of my hand. Being passionately interested in all things natural history, I was introduced to the subject of non-native large cats living in the UK whilst still a teenager, not by design but by circumstance or accident, as I watched badgers during the night... I got more than I bargained for!

This set a president that led me to encounter more unusual goings-on in our countryside and sub-urban areas, I had more encounters and learned a lot about the subject as I indulged in my basic wildlife hobbies, leading to the realisation that this subject must be delved into, and the people educated, and all manner of investigations.

Since 1983, I have been at work, with escalating success in not just investigating this phenomenon but having multiple sightings, finding an inexhaustible supply of field evidence and data, which in turn led me to be well known within the subject. I have given lectures nationwide and been involved from many angles. I have been involved in wildlife conservation for the whole of my life and have worked both professionally and voluntary within these fields, including a 22-year membership with the Bournemouth natural science society and have been the chair of the zoology section for two decades, as well as helping out with curatorship of our museum specimens, president and trustee, lecturing and leading field meetings. I am a taxidermist, teacher and lecturer, philosopher, ufologist, artist, photographer, musician and have many hobbies and interests.

I decided to write a series of books on the subject of wild living large cats in the UK, as I had so much data and believed that knowledge should be available to all, especially something as controversial as this subject. I decided that there was no field guide to the identification of field evidence or guidelines from which people could work, enhancing that knowledge. Also, realising that so many people were in the dark regarding such knowledge, and with so many sceptical minds that refused to acknowledge, I needed to prove a few points.

In this book, I relate my own observations and experiences and relay my findings regarding the numbers, behaviour and species of the animals involved, with many photographs of field evidence. One has to limit the number of photos for publication, and with my thousands of photos, I had to short list the whole lot, perhaps picking the best. The photos in this series are only a fraction of the overall amount of field evidence I have come across. My Studies have been mainly concentrated in my home county of Dorset, but overlap Hampshire and Wiltshire, and I have investigated other areas of the UK.

PREFACE

Why write a book about out of place big cats? Surely there are much more important things to write about?

Yes, there certainly are, and I, for one, would rather be writing about topics that concern humans and the world we live in. The big cat phenomenon is an issue that is close to home, and people can relate to it. Much more serious issues need to be addressed to the public with a softer approach. The issue of big cats in Britain is harsh enough for the time being. Humans are of a delicate nature when exposed to revelations or truths, and this subject in itself creates much horror, debate, criticism, disbelief, awe, and wonder.

This subject is something that I can easily write about as I have first-hand knowledge about it. I believe knowledge should be passed on for all to see, feel, hear or read. Nothing should be swept under the carpet; that just does not work. Too much of human history has been swept under the carpet and important information that has been hidden from mainstream society always has huge backlashes at a later time. If a subject arises that is of a certain amount of controversy, then it should be dealt with there and then, with all honesty, but in a way in which we humans can manage without mass panic.

Knowledge regarding British wildlife has always been a strange affair, and to be straight, we know much about the species of plants, birds, invertebrates, fish, mammals and reptiles, and we depict them in books, their life cycles, distribution and so on, but are we totally right with the data that has been given out? Well, on the whole, yes, but there are still a few tweaks that could be made.

A good example of how people think in the wrong ways regarding animals is the common badger (Meles meles). Most people think that badgers are carnivores and eat meat, and they prey on other animals such as sheep and deer. This could not be further from the truth.

Badgers eat mainly earthworms and other invertebrates, cereal crops, nuts, fruit, bulbs and the odd small mammal or bird. Badgers rarely eat carrion, and most badgers within a colony will turn their noses up at a dead deer or rabbit placed near them. They tend to only eat meat in the harshest of winters when the ground is too hard to dig for worms or in draught periods when the worms go deeper underground.

I have watched badgers since I was a small boy and observed their behaviours. I have baited game cameras with many dead deer and other animals for many years and failed to see badgers polishing off the remains. In fact, only once have I seen badgers tucking into a deer carcass, and in that instance, it was someone else's game cameras in a water flooded area and it was freezing cold.

Most badgers will not hunt animals, but there is often one within a large colony that does actively hunt other small animals, mainly rabbits, mice or voles. Badgers live in large social groups similar to other herbivores. If badgers were carnivores, they would not be able to live in colonies as they would out-compete with each other, there would not be enough animal protein to go around and they would be killing all other wild animals. This just does not happen, and nature would not allow it. But so many British people seem to think that a dead sheep eaten out could be the work of badgers.

Whenever confronted with dead animals, everyone seems to debate whether or not it was dogs, foxes, badgers or big cats. Why the badger? It hardly comes into the equation. Don't get me wrong, some badgers are quite capable of killing small animals and eating them, but it usually does not go any further than wasps, bumble bees and earthworms. Maybe it is because the badger is well known to be strong with a fighting spirit! All very well, but so does a buffalo fight to the death and so does an elephant, but they don't go around eating other animals. In fact, the humble badger is far more of a coward than people give it credit for. It just simply retains some of the typical mustelid traits that other members of the weasel family still have but has lost most of them to become a social omnivore.

If people believe that badgers bring down sheep to eat them, something is clearly very wrong, and this fuels other wrong beliefs about the species. Many so-called wildlife experts seem to think that badgers are top predators! This is very sad and damming. And to think that the species is responsible for spreading bovine tuberculosis to cattle is equally absurd, which fuels the myths. This is just one example among many, but may be the biggest miscarriage of justice and immorality regarding any British wild animal.

The Eurasian badger is not the same as the American badger; this animal is mainly solitary and more carnivorous in nature, within a different taxonomic class. Few parallels can be afforded regarding this species.

Another example is the common rat. Not many people like wild rats, and we have been taught that they are dirty and spread disease and that they are pests to be eradicated in any way whatsoever. Whilst rats are capable of spreading some diseases, most of the diseases they carry have been picked up from man in the first place.

Most rats are clean and don't carry any diseases, even the ones they are most renowned for carrying, such as veils disease (leptospirosis). Some city living rats may carry it, but most country living colonies are clear of the disease.

Humans carry many hundreds of different kinds of diseases and viruses, but rats only carry a few. Rats are also fastidious cleaners and are among the cleanest of all animals. If rats carried many harmful diseases, why haven't we all succumbed to them? Rats are one of the most common mammals anywhere in the world, yet we don't hear of any colony of humans dying or getting ill because rats have spread disease? Yet we all know that rats live everywhere! Hasn't anyone clicked yet? It is all rubbish about them and we would not be alive if it were not for the humble rat as they have been guinea pigs for us in helping us find cures for all kinds of diseases for hundreds of years! We would be knee-deep in our own filth if it were not for rats cleaning up after us, yet as usual, silly humans are not very bright when it comes to changing hundreds of years of old wives' tales and superstition.

So to really understand wild animals and the big cats, we must have a clear head and think rationally about the subject with an open mind and not be tainted with human misconceptions. It is a hard lesson to learn, but unless one and all appreciate these disciplines, then there is no need to proceed with reading the rest of this book. We are dealing with human problems, not animal problems and unless we can see our own faults, who are we to judge? People become brainwashed in many ways and it is passed on to all.

There are also many misconceptions regarding other wild animals, grey squirrels, mice, town pigeons, American mink, crows and magpies and so on. I am not stating that these animals are not pests, but they are not the evil we like to see them as. Remember, it is all selfishness and ignorance, as it is only the animals that are an inconvenience to our modern dirty ways of living that we choose to despise.

When I look at wildlife books and magazines, I am constantly reminded that there is not much in regards to non-native animals except the common types which we all know about. I have looked at several modern field guide books about British wildlife now in the year 2011 and am horrified that in nearly all of them, old data is being used in regards to species and their distribution. For example, the Scottish wildcat only lives in remote Scotland; well, firstly, there is no such thing as a Scottish wildcat! There is a wild cat called the European wildcat (Felis sylvestris), and then there is the Scottish type, F.sylvestris Grampia, otherwise known as the Scottish wildcat.

But what about the European wildcat, the type that also includes this Scottish version? Well, most people do not know about that. This is because so many writers and naturalists fail to mention it. The species lived all over Europe, including all of England, Wales and Ireland, and maybe it still does live in some places other than Scotland. I believe it does. Although there may not be pure specimens living to have hybrids with domestic cats (which were descended from the African wildcat) still means that there were, or are, wild cats still living today either as escaped captives or as relict populations here and there. I write more about this subject later in the book.

The pine martin also only seems to live in Scotland according to many books; now proven to live in many English and Welsh areas. It has taken naturalists (scientists, to be more accurate) over forty years to notice this, as their numbers were dwindling. The general public, many of them competent naturalists, have tried to tell authorities for decades that pine martins still live in areas otherwise not known as dwelling places of this shy animal.

The polecat only lives in Wales, according to some books but has now re-colonised almost all of its former distribution areas across most of England, especially the South West. Again naturalists were not believed by so-called authorities on mammals that they knew existed. I have heard many stories from competent naturalists regarding the same sort of dismissal. Science is often to blame, and armchair academics have no place within this scheme of things.

Another very common and most damming misconception regarding wildlife in Britain is that deer have no natural predators here in the

UK. It is this serious blunder of a myth that helps fuel scepticism.

There is so much rubbish written about wildlife, it is a great shame, but most of all, it is giving across the wrong message. Our scientists, zoologists, biologists, and ecologists have a moral duty to put out the right messages to people, especially if they are looked upon as being an authority on such subjects. People tend to rely on the academic rather than the person taught by the University of Life. This in itself is tricky. I learn more about human psychology in this field than the topic itself, and that is something to dwell on, as it all has something to do with the way we think and go about life. Much can be learned from all angles about ourselves as humans and why we think in particular ways. It may open up the way for a more open-minded way of thinking and gently guide the reader into realizing that not all is what it seems to be, or what it is made out to be, and to search, means to find, and in doing so maybe we can better our knowledge of not just the natural world, but of ourselves as human beings and the relationship between us, our planet, and our spiritual growth.

I have, over the last thirty years or so, gathered much knowledge and information about the big cat phenomenon that I feel that it must be shared, and not just for people's sake but for the animals' sake too. I feel that I would be doing a great disservice not to write a book on this subject and pass on that information, but more so, to put matters straight. People need to know, what was that huge, sleek, shining black beauty that leapt across their garden? Or, what animal killed their sheep? What animal was seen by a coach load of holidaymakers? What animal made that horrid noise during the night? And so on. Questions need answers and this book sets out to answer most of those, but as we now stand, many questions cannot be answered so easily. The topic of big cats in Britain is a phenomenon and has many facets to it, many hidden aspects and strange twists in the tail.

Of course, the phenomenon is not just a silly fable made up by bored country folk in Britain! It is a truly global phenomenon, or to put it bluntly, wherever humans have gone, they have taken their animals along with them and, in particular, Western cultured people. The situation is apparent in Australia, New Zealand, North America, and Europe. Basically, everywhere Westerners have lived and taken their exotic wild animals along with them.

This book has been written in more than one style, the first few chapters being like a novel, and secondly, the more important field signs and my work and the conclusions that I make.

This work is solely my doing and is based on my current knowledge of the subject. I rarely quote other people or use other people's data, as that is not the thing that I do. My work is my work only and it is true experience that leads a person to write a valid account of knowledge regarding any subject. There are a few photos passed on to me by my fellow researchers and some reports from the public. One can compare, but in this situation, there is little comparison. There may be comparisons between certain large cat species within their usual countries of origin and that can be taken into account but I have not based any of my findings on those subjects. There may be comparisons between people's experiences regarding sightings or their investigations or studies, and investigators may find parallels.

Every time I decided that I had finished this series, there would be more data that I felt had to be included, thus prolonging its publication. I had to stop sometime, but it's good that a subject like this evolves and that the people do not give up hope. I also have medical problems to contend with and they have hindered me for twenty years putting great strain on my fieldwork as well as writing. Had I not suffered fibromyalgia syndrome combined with nerve and back problems and possible Lyme disease, then I may not just have had this book completed years ago but also may have had the smoking gun evidence in the form of photographs, video or biological evidence.

All across Britain, from the far North of Scotland to the south coast of England, to the hills of Wales, to the flats of eastern England; From the mountains to the sea, and from the plains to the cities. In fact, all across the isles have been reports of large cats mainly since the nineteen sixties.

Britain is not alone, no! Australia, New Zealand, North America and mainland Europe all seem to have many reports of out of place large cats. The reports from Europe are more recent within the last twenty years but mainly in the last ten years. Evidence from Australia and New Zealand goes back possibly hundreds of years, but more so in Australia, where reports surface from the Victorian times, very similar to Britain. In North America, pumas exist naturally and the jaguar lives in the most southwestern states. At one time, their distribution went further north. Many jaguars are black naturally, so many people in the USA think that big black cats are melanistic jaguars, but in reality, they are more likely to be black leopards, and some of them could be a black puma, especially if they have been reported from the East. (I will explain later).

This book is basically an educational tool for all people interested in truths and the nature of our earth and of humanity, and for those people wishing to know the truths about alleged large cats living wild in the UK. This book is for the converted and the sceptic. This book is for the ecologist and the biologist. This book is for the conspiracy theorists, the realist and just about anyone, as there are many messages hidden within, messages that subconsciously train one's thought. It is also in everybody's interests to know what wild animals we have and to take a responsible approach to the phenomena; even with its small health and safety queries, we all should be told the truth in a way that does the least damage to the sensitive fabric of human mindset.

I may repeat myself or go over the same ground twice or thrice within this book, but that is needed as it takes several attempts to get the knowledge out into the most receptive of human minds. You can tell a person a dozen times and still, they will ask the same questions.

1

MY RESEARCH

Although I had been doing my own research from a wildlife perspective for many years after my initial sightings in the 1980s, I decided to do a more serious take on the matter after the year 1995. I could not work much because of ongoing illness in the form of fibromyalgia and back problems. I decided that I could sit around doing nothing and had to do something worthwhile and for the rest of society. I worked low key on a part time scale or as much as my body allowed. I could not walk far often, so my chances of finding what I was looking for were going to be very difficult.

Initially, the places where I found large cat evidence were not the places where I had the most witness reports. Wherever I visited places where people had reported large cats, the chances were that I would not find much in the way of evidence. Where animal deaths had occurred, then there was much more to go on. As time went on, I had better reports not just from anywhere and from all over the country but from some of the key areas where I had found evidence, and these areas were also high in deer numbers. The more sightings I got, the more I could fit them in on a map and access the areas relating to other environmental factors needed or not by large cats.

Eventually, I was getting more reports from better areas, and the areas where I concluded there to be breeding leopard or puma seemed to get more of the better reports from country living people or people on holiday.

I started to get many eyewitness accounts after the public was aware of my interest in the subject, and people started to know about my research, often via my website. More so, the witnesses themselves were desperate to tell somebody about their incredible sightings. Many sightings would go to other large cat researchers or to the police or wildlife charities, who, to be honest, had no right to deal with such reports because they did not even believe in the existence of large cats. Some of these bodies passed on reports to me, but most of the time, they did nothing about it and certainly did not help the witness in any way, but often disappointed them by stating that they must have seen dogs or other wild animals anything but large wild cats.

People are, so awestruck by their sighting; it leaves them dumbfounded; literally in mild shock in many cases. This is in itself proof that what the person or people saw was not just a common animal but something very different, enough to change their whole psychological perspective.

In the vast majority of sightings, the witness really does know what he or she is looking at and also the size. People are used to seeing wild animals and I think it is so wrong to assume that the person had seen a dog, fox, badger or deer when most people are so used to seeing such wild animals, if not in the flesh, then by pictures, television, media etc.

The profile of a large cat is obvious in most cases and the witness at once notices. Of course, there is another element to this and that is the fact that many people want to see a big cat or somebody is just silly and naive and suggests that the animal they saw was a big cat.

There are certain types of people and with different behavioural characteristics, after years of observing them, there can even be stereotypical behaviour from certain people from all of the many types. I touched on this subject in the last chapter, so the reader will know where I am coming from here, but certainly, if I needed to research the subject seriously, I had to know what people were being very honest about their accounts. I dismissed the namby-pamby person, who had a touch of madness about him or herself, and there are many like that, people who were just looking for attention and concentrated on the folk that were intelligent. Sometimes though, it is these kinds of people that see cats often, and sometimes there is large cat field evidence from the areas in which they have had sightings or other happenings.

I also concentrated on country living folk, as the areas in which I chose to study were indeed countryside or just bordering towns or villages. As I was familiar with all parts of the south coast, from Dorset and the West Hampshire New Forest and the borders of Wiltshire to the north, I decided to make five study areas. These areas were areas of multiple sightings and great wildlife habitat. The areas were also very big. I have included the map that shows the basic areas.

The problem is, is that I do not want to be too particular in regards to places or areas where I have found much evidence because there is the possibility that some folk will use the information to their advantage. In some respects, I am not too bothered because I know how difficult it is to track these animals and if people were intent of harming them, then they would have been extremely lucky, or they would need to know a lot about the beasts and their habits to be of any threat.

It would be very difficult for anyone to hunt one of these creatures down. Even researchers have had enough time just looking for field evidence, let alone shoot a cat that could be hiding anywhere in a thick forest and heathland within a twenty-mile circumference! In effect, one has already proven to a point how difficult it is because many other researchers have trodden all over my study areas where I have seen large cats and found all kinds of evidence, they have even placed hundreds of cameras and as of yet, only got one or two fleeting snaps in many years.

The local press has given me good leads as to where sightings have been and like many local newspapers all across the UK. Sensationalised articles about the certain beast of wherever appear in clusters throughout the year and are a good indicator as to what cats may be doing. Followed up interviews with the witness are also important but not crucial. The more witness accounts, the better. One has to get an overall picture. I have still only got many fragments of the picture at this time of writing. I have found it so difficult. The areas are so big, the animals so different, the witness statements vary, and the cats seem to be on the move a lot of the time. However, over the last ten years, I have built-up a fragmented picture of something. It resembles a five thousand-piece jigsaw puzzle with only a few hundred scattered pieces, some joined together in clumps, some joined by long arms, others just one piece in an empty sea. The missing four thousand pieces are out there somewhere. Embedded within the thousands of acres of heathland and forest, farmland and hills.

If I were making a study of the cats' main prey species, such as the deer, then it would be relatively easy. But hang on a moment; the deer! Of course, the deer. To study the cats, one has to study the deer. Easy as pie? No, sorry, it is not that easy, but certainly, one major part of the puzzle is to study the deer. I have found that the deer do indeed play a huge role in the subject and without them, things would be totally different. In fact, it is clear to say that if my areas were not so full of deer in the first place, then perhaps large cats would not be so prevalent. It seemed logical to conclude that the deer were the main part of the puzzle as the cats' workings revolved around the several deer species. Although initially, it seemed logical, as time went on, I seemed to find out more and more simply by following the deer movements.

In two of the areas in which I chose, there were huge herds of wild sika deer. I will number the areas here from one to five. Numbers one and two are large areas of great biodiversity in south Dorset and they are my key areas. Both these areas are just a short distance from one another but also have no true defining borders. Area three is about a couple of miles from the other two but has definite barriers such as roads and rivers, but is a large area of heathland and mainly forest, of which is mainly conifer plantation. This area also has large populations of wild sika deer. These three areas, although large, are small in comparison to the remaining two. One area is the whole of the new forest, a massive area. And last but not least is an area of which comprise farmland, chalk downland and much woodland. This area straddles the three counties' borders of Dorset, Hampshire and Wiltshire and includes some of the finest wildlife habitats in the country, including a large national nature reserve. There are no sika deer but instead many roe and small populations of fallow deer. These are the main areas in which I purposefully set out to gather data on a regular basis. Also, there are many other local areas in which, over the years, I have made small preliminary studies, but not on a regular basis as to be able to include them in the main study areas.

At times I now may get three reports in a week. These are from all over the UK. Nationally there may be upwards of four thousand reports a year from all areas to all authorities or researchers. Many reports are of multiple sightings, of mother and cubs, of a large cat carrying a deer, a fox, badger or a domestic dog. There are so many sightings in all different contexts that it would be fruitless to include them in this book.

Other authors have done grand jobs writing books of people's sightings. They are all amazing and sceptics should read these books and learn a thing or two! Most reports are genuine! Most reports are really of large out of place cats; otherwise, the person or people would not go to great length to report it in the first place, especially when they know that they will be ridiculed. Most reports are from adults rather than children, as adults are more responsible and knowing! That is not to say that children don't report as they do but more often with their parents. Many children have seen large cats whilst out playing or cycling and reported large cats seen to their parents but not believed. Many children have come face to face with large cats, even mauled by them and then they are believed by parents but not maybe the public.

Witness reports are not just of a sighting or two or three but actual encounters in many different ways. Many people never report, but maybe an inner circle of friends or family, so there are many unknown areas of sightings and other close encounters. Many farmers and landowners do not say a word but try to eliminate or actually shoot the animals and say nothing. Many people see large cats and think nothing of it, thinking that it is normal or nothing to write home about. Many people see large cats but don't think anything about it because they are not curious or have no interest whatsoever, even if the animal runs across the road in front of their car! Other people, such as the gun-happy morons, see them often and shoot at them without saying a word, but often they admit when one speaks to them alone. People have touched large cats by accident, people have been struck by large cats whilst defending their prey, and these are usually known about, especially if they sustain injuries.

Multiple witnesses see them from buses and trains, and groups of hikers, and birdwatchers see them also. Holidaymakers from abroad see them. People holidaying on campsites see them often. There are so many people seeing these large cats all over the UK. No person can ever say just on these happenings that they do not exist, let alone the field evidence and the scientific proof.

The dozens of other researchers from across the UK all get dozens or hundreds of reports also, and now most of them are being filed on a database. The police maybe receive more sightings than any other authority, but of course, they are rarely passed on to researchers. The code of conduct here is to cover up and tell lies! It is ironic to know that the police have more data than anyone on witness sightings and reports and also have some of the best video footage anywhere of large cats. Their helicopters with thermal image technology have had many a large cat in their sights and recorded. The military has many good reports and footage using night vision equipment. These again are kept secret. But, with maybe only one in a dozen sightings actually reported, maybe we will never get the whole picture unless people open up and the subject gains in popularity and acceptance. The reports generally are from all areas. There is not one town, village or hamlet in the whole of the UK mainland that has not had some sort of sighting. Most areas will have had hundreds over thirty or fifty years or so. There are hotspots, though, but we must not assume that because one area has more sightings than another, that one area has fewer cats than another.

For example, a young male puma left its parental area to search for its own territory. Having passed through two areas owned by other male pumas, it had to keep roaming. It zigzagged across thirty miles of countryside, passing twenty-five villages and bordering two large towns, a railway line and a motorway. The cat was desperate and decided to roam during the day (as many a large cat does, that is on a purposeful mission!) To avoid confrontation with the dominant males, and so was seen by many people from different areas during day and night. People may have assumed there to be several large cats and each person's description of it may differ, leading to the belief that different colours mean different individuals. Whilst this is often the case, many mistakes can be made and indeed, one animal could be responsible for hundreds of sightings. On the other hand, people have assumed that sightings in another area were of one particular individual but, in fact, were five separate individuals! Especially if there are cubs involved, a mother and a male.

The sightings I myself collect and those from all other researchers are very important for many reasons, and the witnesses are often content that they have been believed and their sightings recorded.

I myself do not use people's sightings as the basis of my work. I don't need them, but they are handy, and sometimes they cement suspicions and lead to other evidence. One can get a good idea of an animal's movements if sightings are consistent. A found sheep or deer kill could be attributed to leopard number one or two, for example, and one can assume that a cat's area has enlarged or shrunk, or maybe it's a male or female etc. The difficulty now is that there are so many large cats living and breeding in the UK, there are so many possibilities. With dozens of dispersing juveniles every year roaming around searching for a vacant place, then there are going to be many more sightings, especially if numbers have reached saturation point, then many animals will constantly be on the move. This is also a semi-natural biological normality for many species but especially predators. So-called satellite males may be frequent animals that have no given territories but roam in search of females and may or may not meet the males that have territories overlapping.

Where Are The Large Cats?

Rural living

Large cats have been seen almost everywhere across the UK. There is no place that has not had its fair share of reports. There are areas that have far more reports than other areas, and there do seem to be hotspots. There also will be areas where cats are not seen very often, this does not mean that there are no cats about; we must remember that we are dealing with animals that are rarely seen by humans, so for somebody to have a sighting in the first place is an abnormality. When thousands of people throughout the Uk see large cats every year, then obviously there have to be many individual animals about. They will be roaming around, some having established their territories, others not so but may be looking or passing through areas. There will be areas that are more suited to certain species, and these areas may not get reports from the public. Leopards and pumas are renowned for living amongst people without being seen. They certainly do in Africa and Asia, even in cities, these animals live and are occasionally seen by people.

In America, recent studies have shown that puma will live on the edge of huge cities also but rarely seen by people. To start with, in Britain, we must follow the witness reports to get any decent idea as to where the cats are, and this over the last forty years has yielded a basic map. There are areas where there would seem to be more animals living and areas where few are living. I often am asked by people, "Where should I go to see a big cat?" The answer is nowhere and everywhere. Cats are seen mainly by accident, and they are often in areas where one would not expect to see them. So in areas where cats are known to live, there is possibly much cover and the animals have home ranges and know-how to keep a low profile, so actually, the areas in which cats live most may actually be areas of fewer sightings than areas where roaming cats are forced out into the open and seen by people.

Cats are secretive and know how to hide and keep a low profile, from domestic cats to tigers; they can all seem to follow a basic set of rules that is generally cat orientated. Even a house moggy that would normally be lap friendly and constantly walking under its owner's toes within the house will change into a shy wild creature when out walking and stalking the fields and hedgerows, and owners often see his or her cat run away from them if it is out in the fields. This just goes to show how the cat can adapt.

There have been sightings of leopard-like cats on islands also. The Isle of Wight has had quite a few sightings and maybe an owner or two have admitted releasing them. I heard a story about a cat enthusiast who released a leopard. Most of the reports were of a black panther, video footage from a mobile phone showed to me what looked like a leopard. Although the cat could revert back to suddenly being wild and elusive, which may be a natural trigger, it is still more likely to be seen by people and followed until captured. The island is large but not too big that one would not hear reports from somewhere else within a short time. There is the possibility of more than one panther living on the island. The owners may have released two, or someone else may have released one, or another escape may have happened, given the fact that the island is small and there not likely to be more than one keeper of exotic large cats other than the zoo. It is more likely to be also found because it would be more likely to feed on domestic animals.

There are no wild deer on the island, or if there are, then they are so few in number to sustain a large cat for a few weeks. Deer have been on the island, fallows have been sighted, muntjack has lived for a short while, and red deer also have had a short wild time but no roe. All the known escaped deer from parks and farms manage not to survive. There is a good rabbit population and fox and badger. Without deer to sustain an individual, they would have targeted fox and badger populations more and it would have been noticed. As yet, there are no reports to me or anyone else I know of that has had reports of eaten farm animals. If there are large cats still living on the island, they must be extremely elusive and feed on smaller mammals and birds. If there were breeding of any kind, there would be more reports. It is very different from the mainland. There are, without a doubt, though, reports of large brown cats looking like puma during the 1980s and small jungle cats were killed and sighted.

The Isle of Portland in Dorset is joined to the mainland by a causeway, it has been a place of many a sighting of four different species of cat. I have found much evidence myself for at least two species. The many quarries provide safe birthing places and rabbits and sea birds provide food. I do not think that cats spend long amounts of time on the isle but short breeding breaks or accidental entrapments or wonderings. There is a long strip of land joining the mainland, but it is open also. There are at least two ditches which could conceal a slinking cat, especially at night. The other side comprises of the chessel bank, a huge long section of shingle, hard walking but worth it for a cat that needs safe, deep caves to house cubs in for a few months.

The isle of Anglesey in North Wales has many sightings of leopards. One has to wonder how the individual got there; had it swam or walked the railway or road bridges? Well, it could be any one of them; the fact is that there is at least one leopard-like large cat living on the island with rather low biodiversity.

The Isle of Sheppey has many reports of large cats, and I have seen evidence of large cats being on the island by another researcher. Sheppey is not so much an island, more like a peninsular with road access; cats can cross bridges at night. To the cat, it is not an island! A cat can walk along long bridges and railway lines onto any island with ease. The Isle of Sky has also had large cat sightings, Mull, Isle of Man and of course, Ireland, the republic and Northern Ireland.

Portland in Dorset is a peninsular joined to the mainland by a long causeway, but amazingly, the island boasts big cat history. It has a long causeway joining to the mainland and it would not be difficult for a determined cat to take an hour or two to walk it, hiding in the ditch that covers large areas on the eastern side. The western side is the shingle of chessil bank. A cat could walk all along the beach from further up out of town and end up under the western cliffs. Portland has had many a sighting from local people. Oddly enough, all the large cat species have been recorded.

Jungle cats have been recorded living in Weymouth. There are huge reed beds around Radipole and Lodmoor lakes. Black panthers have been spotted crossing main roads in and out of the town Leopards have been seen peering into a badger sett.

Roe deer carcasses periodically mount up and dogs go berserk in the night. I have found much good field evidence from all around the Weymouth areas, including Portland.

I have found leopard-like and lynx-like footprints on a few occasions. One time whilst out with two of my friends, Darren Naish and Mark north, we found the tracks of possibly two species in just a few hours, along with many scats and a possible den.

The island is full of quarries and rabbits. It is the ideal denning site with much food for the first few months of a cat's life. I am not surprised that cats of possibly all three species have often used it for this purpose. There are many quarries within the county and large cats may travel distances to breed in them, or certainly some of them if it isn't already on a rival female's territory.

The Isle of Skye also has had reports. There is not a lot of cover on Scottish highland isles but often many red deer. Undoubtedly there will be some large cats visiting them, maybe even swimming to some and back. There is a naturalised population of all three species in Scotland.

Wales certainly has far more sightings than Scotland at the moment. That does not mean anything particularly: Wales is full of wooded mountains, lots of cover, a growing population of wild roe deer and muntjack now, many badgers, hares, rabbits and all other wildlife. Feral goats live on many coastal cliffs. Wales has so many quarries and rock formations, ideal denning places all over the country. Scotland is far more remote and many of the cats will be living in these areas where humans are sparsely scattered.

The red deer in Scottish highlands would be key for survival in the winter months, with other common species being eaten also. Roe also abounds and hares in the mountains. Feral goats abound in some areas. On Northern Island, there could also be breeding leopards. There were many kept in the North and the Republic, and new licensing did not come into force until much later than Britain. The island has its own researchers and much evidence in the way of photos, footage, and livestock deaths. Ireland does not boast as many species as the mainland does, but feral populations of deer could be sufficient in some areas. The country is big, with a big wild landscape, but there are not the same numbers of reports that come from Britain. There could be breeding as there have been reports of possible cubs. Most of the cats are recorded as being black.

There will be far more large cats living in rural areas than suburban or urban areas at the moment if they are not being persecuted hard. Cats ideally prefer large areas to roam with a variety of game and without hunting pressures or disturbance from people or dogs. This may not be the case, though, as other things determine where any individual cat lives. Availability of vacant territory plays a vital role and can actually mean all the difference in regards to whether a cat lives in woods or fields or the middle of a city. The country cats may not necessarily be the fittest or healthiest individuals, but they may find it easier to find mates without the chances of being killed by traffic but may have more problems with shooting or hunting. One would expect city cats to be smaller than rural ones, which would most likely be preying on larger species such as deer rather than rats, foxes and dogs.

Corridors

Throughout the last twenty years or so of many thousands of large cat-related reports, there is a picture that has formed that seems to suggest that the majority of observations had taken place in certain areas that have certain features such as railway lines, golf courses, sports playing fields and industrial sites. There is an obvious reason why large cats have been seen in these areas, especially railway lines, as they are routes to and fro. Areas where humans never go, places where the animals feel safe to move around despite noisy trains and electrified rails!

I am sure that one or two cats at least have succumbed to this lethal killer, if not the fast-moving train, then certainly electrified on the rails. These railway lines are a vital link into the distribution success of both leopard and puma. The cats use them day and night and are either use them as rat runs or they encompass cat territories. They are often used as territorial borders, but many a roaming cat will find themselves on a line and run, going through built-up areas, including cities.

Cats are searching for mates or new territories, especially juveniles, who will use railway lines to get out and about. Some of these lines are habitually used to get through built-up areas by the cats. It also opens up the gateway for cats to be urbanized. Many of these railway lines, whether they are used or not, will run for thirty, fifty or over a hundred miles. Roads will usually go underneath them or over them, lessening the chances of cats having to cross roads especially busy ones, so thus lessening the possibility of putting their lives at risk. It is this reason alone that aids a small number of cats to eventually find each other over large distances.

Golf courses are also attractive as many are not open to the public, and not so many dog walkers, whilst some will attract leopards for the very same reasons. At night they are not used and there is little hunting and shooting on golf courses. They are also good places to hunt rabbits. In some built-up areas, they may be the only green areas and so may be central to a cat's territory if it is a city living animal.

The same can be said of churchyards and graveyards for the same reasons. In areas where all of these occur, especially on the edge of towns where there is ample food, leopard and puma will be using the area.

Within the deep countryside, even the country areas that have large towns within valleys, the wooded slopes of the hills will be home to large cats. There are many such places within Britain where the woodland sloped down right to the town edges and it is from these regions that we get many reports of sightings or of deer carcasses and cat-like sounds.

Sea cliffs and coastlines are also other corridor areas for cats to travel, and a cat that knows any area may use these areas. In Bournemouth, near to where I am based in Dorset, large cats have used the six miles of sea cliffs to move from one area to another without risking crossing roads or meeting with people. A short route from Purbeck through Poole to the New forest would be for the cat to travel the railway line straight through! Or to use sections of line to get into Poole, then to the coastal cliffs and just amble along all the way to Hengestbury head, where the cat can turn off into nature reserves on the edge of the New Forest. A cat could easily do that in one night.

I had once seen puma footprints along the beaches and clifftops suggesting that at least one animal actually had done just this and the chances of others doing the same are very likely. Even deer accidentally find themselves in Bournemouth or Poole centres because they have come off the railway lines at the wrong places and get lost, or find themselves at the sea cliffs and then find their way back out by following the obvious linear routes out into the forest again. Cats would do this by a mixture of accident or purpose. The railway line then becomes central to many cats' travel routes.

Canal banks, river banks and river valleys are bottleneck areas and funnel travelling cats through built-up areas. These are the areas to put out trigger cameras. When looked at from the air, any area can show up possible travel routes for large cats. Google earth is excellent for sussing out possible routes through most UK cities or large built-up areas. Some cats may never need to travel far and spend their whole lives in the countryside, not seeing many people or even having anything to do with humans. Most cats will see and hear people on a daily or nightly basis, either from a stationary point or as they travel their beat. Some cats may spend most of their time in thick forests, others in open fields. Some will be in high mountains, others in low hills. Some will be in river estuaries, others in city centres. Some will be living in huge industrial estates behind fences during the day and car parks at night. Many wild pumas and leopards, visit zoos and wildlife parks that house the same species or similar. Most large cat keepers at zoos that are in the countryside tell stories of the wild cats that come in when the puma is on heat, or when the female leopard keeps calling, etc, etc. They either see them or find their field signs or catch them on security cameras.

City slickers

Leopards are well known for living in towns and cities in India, to a lesser extent in Africa and Indonesia. In some parts of India, leopards live at high densities, which would not have been believed if it were not for serious research carried out by certain naturalists or researchers. If all the requirements are met, then a large cat does not need to have a huge territory and this would mimic foxes in that respect. There are many more reports of cats living in towns and cities than there were twenty years ago. In Britain, it will possibly be a similar situation as the countryside becomes smaller due to lack of free territory or because of human developments. Leopards and, to a smaller extent, pumas do adapt to these situations, possibly under much stress at first. When several generations of city animals have been bred, then it is more likely that they will continue the trend without needing or wanting to return to a rural lifestyle and individuals may become smaller. There are more suburban cats than true urban ones, but who knows? Maybe there are leopards living in the heart of London! There are certainly some living on the outskirts and also in Epping forest areas and the M25 motorway!

The areas around Epping and the well-known forest have huge herds of fallow deer, and always have done. In more recent decades, muntjack has been added to these natural cat food species. There have been many reports of some very good encounters of close up sightings regarding mainly big black cats but also just a few brown individuals. The North London suburbs do not constitute the most ideal of habits, but a slight lack in game rearing or shooting in some areas along with such high amounts of deer, offer some of the best habitats. Corridors along streams and areas where no footpaths criss-cross are ideal. The amount of deer is enough to sustain a breeding female or two and a male or two. To this day, huge herds of over a hundred can be seen from the main roads in some areas and the herds one can see are the tip of the iceberg! With such an abundance that is not as managed in the same way as they are in other deer areas offers good breeding grounds. As I amended this section in 2019, a report came in regarding a big black cat seen at a small nature reserve in North London. To date, I have heard of dozens of reports from this same area. I have reports from the middle of Bristol, Ipswich and Liverpool and Glasgow.

I will write more about the local city slickers later. Most large cats will be living in the countryside, perhaps eighty per cent at a guess, but this may change in the future if they are persecuted as the cities will be their only refuge and be very difficult to eradicate. In some parts of North America and Canada, there are more pumas living in town areas than in the natural countryside, this is a direct result of persecution. Large cats may hate mankind, but sometimes they may learn that the best way of living is to live among them. If there is more food available, then it makes sense, but only under pressure as surely no self-respecting leopard or puma, the status symbols of wilderness and elusiveness, would choose it if they had more natural options!. It may be that many city-dwelling large cats do so because they are the overspill of an already saturated countryside! Food for thought, yes, but could there really be that many large cats living and breeding within the UK?

2

THE FIELD SIGNS

One of the main topics within the realm of big cat investigating is searching for field evidence. Every animal leaves clues to its whereabouts, what it has done, what it may do, how healthy it is, whether or not it is ready to mate etc. It is something that any tracker, whether hunter, photographer, movie maker or just general naturalist, needs to know and it is partly instinctive to look for the field signs of any given animal.

When one opens one's mind to the available data that seems to be everywhere when one is looking for it, one realizes how all the animals, especially mammals, have their own special networkings of communication and that the whole mammal system is based on visual and olfactory communications and is very similar to our computer worldwide internet and Facebook. Everyone knows each other's business. We humans do not use such communication systems and rely on computer technology to do it for us, living outside raw nature; we tend to forget the possibilities of animals and what they do. A new world opens up to the apprentice.

The various field signs made by the different cat species are very similar to each other and it can often be impossible to differentiate between the species and what marks they have made. Within modern writings about the communicable behaviour of leopards and pumas, there are differences, but then later, that data may be disregarded or changed. Female pumas are said not to scrape or spray. In another report, it states that they do not spray but scrape, in another, it states the puma male doesn't even spray but makes lots more scent scrapes than a female. The latter, I am more inclined to believe. Most of the stinking sprays in sniff out seem to be accompanied by leopard spoor, scats or scrapes in areas where I know that species to be more active rather than pumas and in the areas where I am certain there are pumas using the areas more than leopards, I find no stinging sprays and only a few very large scent scrapes as if done by large male pumas.

What makes it easier are the different types of marks made in a certain area; one can assume that it has been done by one individual or an individual's family group, or a female with an attendant male waiting to mate. If the signs persist for long periods of time, such as nine months, for example then, then one could base one's findings on a possible territory and then everything found within that certain area could, in effect, be the doings of that individual or family group.

If there were a typical dividing line in relation to territorial marking, then other signs within the area could be attributed to either party. When I first started trekking (for want of a better word), I missed much evidence from the forest floor. Cats make more visual signs than any other mammal.

Deer follow in a close second. They share a subliminal connection to each other in at least two different ways. The first is that their markings may be very similar to each other that it can be difficult to differentiate between deer and cat secondly the two different mammals may, to some extent, need to understand each other's signs, as from the cats' point of view, they need to know what the deer are doing to keep a food source within bounds and from the deer's point of view, they may need to know when the cat is back in the area and hungry.

When one finds a cat territory or part of it, one can expect to find field signs on a regular basis. At certain times a cat may keep a lower profile than usual such as a female with cubs or a young wandering male. At other times older well, established territories may show up on many field signs. Whatever way one looks at it, the field evidence is the best way to establish whether or not a large cat is active or not in any area, what species it is and what species it is eating, and the collection of material for genetic research opens up larger areas for research. People these days have lost their way and rely too much on photographic evidence rather than field evidence. This is very sad and a constant reminder of just how far humans have detached themselves from nature. Photographs can be manipulated in a huge way, but a pug mark is simply nature's proof full stop. I will write more lengthy data on the various disciplines of field evidence in separate chapters later.

A typical leopard scrape from one of the Avon leopards, possibly one of the two females.

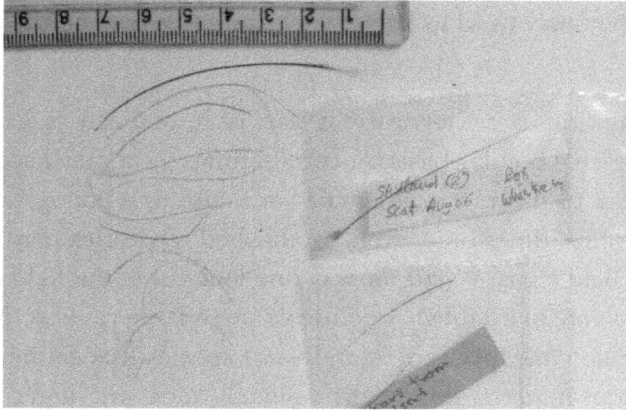

Cats lose their hair in many ways, and moulting is the most obvious. Barbed wire fences claim much of this if the animal is moulting or not, but especially if it is in moult. Usually, it is the back and tail hair if the animal goes under the whole fence, or it can be the belly and back and tail if it decides to go through the fence (more common with deer)or if it leaps the whole fence (unusual) then some belly hair may get caught. Rarely have I found whiskers caught in the barbs. In one area where I searched for evidence in West Dorset, there was a two and a half foot gap under an area of wire fence, yet there were several thick whiskers twisted around the fence. They were wrapped around three times which meant that the animal had to of somersaulted three times over the fence to try and free it. It did leave half of its facial whiskers deeply implanted for me to find. As soon as I found these, I knew what they were. One of them was three inches in length without tapering, which meant that another two or three inches could have been added to it, and they were brown with white bases for about an inch. The only animal that has bi-coloured whiskers on a regular basis is from cats. Foxes have jet black whiskers and they taper quite far up, rabbits and hares often have black and white whiskers, but these are small. Deer do not have the same kind of whiskers and they are not firmly planted and come out easily. Wild boars also have a few short whiskers that are easily removed. Dogs rarely have long whiskers. The only explanation I had was that they were whiskers from a large cat. I took them home and under the microscope, they did indeed show the medullar pattern of a large cat-like a leopard. I sent the best one away for a TV documentary, monster quests (The beast of Bodmin) and it was examined by an expert. Strangely enough, the verdict of that whisker, along with several other bags of hair, was every other wild animal but cats. Oh, and fox. This made me highly suspicious as if it was not a cat, then it could have only been a fox and a very large odd coloured one at that! There are professionals making fools out of people like myself, in a desperate bid to know the truth, they are hiding the evidence. It has happened to many large cat researchers. Some people have

has happened to many large cat researchers. Some people have
even planted genuine large cat hair from captive animals and
sent it off for genetic analysis only to have it return as a domestic
cat or dog. I do not think that it is pure incompetence on their
part but a genuine conspiracy to hide the truth. Another whisker
from the same batch of hair was sent to a university for analysis
only to come back as being from a badger! And the technicians
said that they were sure that there was no contamination.
Badgers have small whiskers as they are constantly being
broken off as it is a burrowing and rummaging animals. All
badgers have small whiskers; usually, they are not visible unless
one is at close range of about a meter or two. I have skinned
many badgers within my lifetime and know all too well that
they are an exception in regards to having visible whiskers. I
have never seen a badger with long whiskers, but I know that
some do have whiskers of up to three inches long as there are
pictures of one or two out of hundreds that do, but they are cubs
or animals kept in captivity. The whiskers in question are way
too big to have come from a badger unless it was some sort of a
mutant. Also, the whiskers are very thin and wispy.

A leopard scrape from the three counties female.

Leopard scrape.

Scrape with taken soil samples and hair spossitive.

Scrape at the three counties borders.

Fence posts are often used by large cats, especially if they are a feature within a large, mainly featureless area, such as open heathland. This large post is placed on a low hill within a large area of moorland and is just off a game trail. Every large cat that comes this way will smell or rub against it. There are many black hairs and small two-tone whiskers embedded in the splinters of wood. The hairs are approximately three feet from the ground. Looks less glossy. Puma hair can be long, also lynx, especially in winter pelage. The hair from the belly area of most cats can be very long, several inches in lynx and in the case of the lynx, the ruff hair can be long also. The tail hair of most species can be long also but not as long as the belly or neck hair. Dog hair is usually matt and soft but with a hint of wiriness in it. It can be any length.

This is a barbed wire fence with typical hair embedded within the barbs. By placing one's hand along it, one can easily see the hair against the near background of the hand.

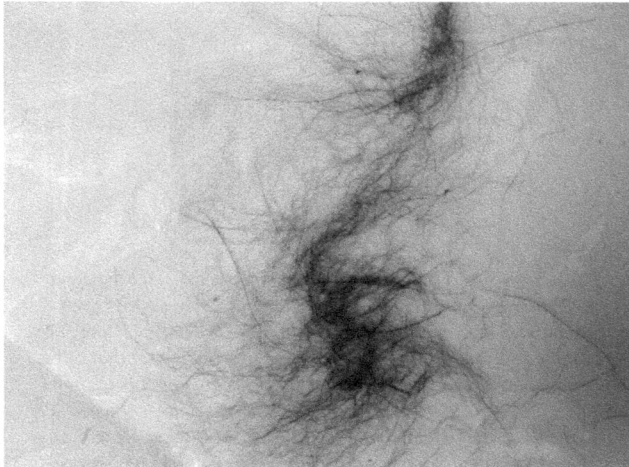

The clump of hair from barbed wire fence showing both underfur and outer guard hairs of a black leopard.

When leopards go up and down trees, they usually leave claw marks and when a tree is often used, one can see the same claw marks as old and new. Here the old and new are in the bark of a large Scots pine tree with horizontal boughs, ideal for watching and catching prey that has been taken up into the tree.

Plugs of oak bark are removed when leopards regularly shin up the trunks.

Claw scratching is uncommon to find and may be done by all the cat species, usually to sharpen the claw ends or to get rid of burs on them. Claw scratching is often used as territorial marking, especially if it is done on a prominent tree by the side of a path. Both leopards and pumas will do this and other cats may mark the same tree. Cats may also mark the trees used by roe deer and often, it can be hard to separate the two species. Roe deer bucks are very territorial and will mark the lower sections of usually small trees with their antlers as a visual marker. They score the smooth bark and rub scent on them, usually with a patch of scraped earth underneath the tree on the ground. Hoof marks can be present and urine or musk. Deer also rub trees to release velvet from hardening antlers and usually, more bark damage is apparent with remnants of velvet.

A well used scratching tree near Corfe castle.

A puma was visiting a business area near Fordingbridge in west Hampshire, within one of my study areas. She often visited a stable and left hair and muddy footprints on the fence as she leapt it. One of the horses got ill with colic after being afraid of the cat. The hair I collected was very white and looked like typical puma hair under the microscope.

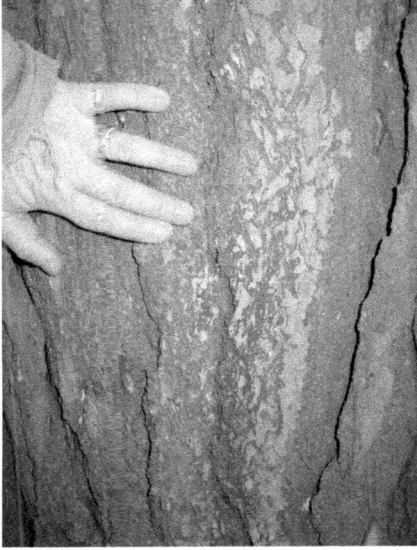

A closer look at some of the launch off trauma.

A closer look at some of the launch off trauma.

One of many scent scrapes within Meyrick Park in the centre of Bournemouth. Ironically three different individual leopards were using the area in a single year. There had been a female for several years using the park and other areas of the Bournemouth conurbation, but in 2018, a male left his footprints and another animal, most likely another female, became very territorial against another. Dozens of scent scrapes were deposited along trackways and in woodland runs.

There is a special place where the Avon leopards feel safe enough to eat their kills at leisure and often take carcasses into big oak trees. They often remain there to be dislodged by high winds. Below this oak are the remains of four animals, two roe deer, a calf and a badger.

The Author finds a cat lying up in the area under bracken on heathland.

33

Most leopards and pumas have a habitual safe feeding place (H.S.F.P). They can drag or carry carcasses over long distances just to feed at ease in such places, so bones pile up. This site had the remains of sixteen sika deer carcasses courtesy of the old male leopard from the growling corner.

3

IT GETS HOT: MY LATER SIGHTINGS

The London Leopard

I was on a train bound for London as I did several times a year during the early 2000s. I always made sure that I had a window seat and sat face forward to travel rather than seeing things behind me. I was always looking for what wildlife I could see. There were about fifteen to twenty people in the carriage I was in and most people were quiet and busy on their laptops or mobile phones, taking no notice at all of the real outside world. I always took particular attention to the areas below London around surrey and home counties areas where there were huge areas of woodland without many farms or villages for miles. This was a traditional puma country for years.

On this day, it was cloudy in late September, and the train had been quite slowly moving due to a hold up on the line. We had just started moving off from a standstill at Esher. As we passed this place and station field came into view, then suddenly a huge jet black beast was trotting in quite a hurried gait towards the railway line. It was either a big male leopard or a jaguar. It had a huge massive head with light green eyes that seemed small within such a large, broad head. It had a massive muscle body but not as thickset as a jaguar, so I assumed it to be a huge mature male leopard. It was just about to leap over the high fence that bordered the line with the fields and I must have had only three seconds of this sighting, but it was one of my closest as it was only a couple of meters away yet behind the glass windows of the train. It was right there in my face, so near I could have counted every whisker on its cheeks had I time to do so, so beautiful was this animal. The more striking aspect of it was that when we passed the hedgerow and the field alongside was presented, a man with a rifle over his shoulder ready to use and two spaniels at his side were marching down the field toward this leopard! I just hoped that it was a coincidence rather than a purposeful attempt to kill it. The cat was in a hurry, whatever the situation.

In a flash, the image went and I quickly looked around to see if anyone else was looking out the window. They were not, so I didn't even bother to ask anyone as they all seemed to be oblivious! I reluctantly told my companion who was travelling with me if he saw it, but I knew he didn't as he wasn't even looking out either. I don't think he even believed me when I told him what I had just seen.

There is a typical domestic dog footprint, and this is it. There is no doubt about prints like this. The overall shape is oblong in both front and back feet. The two top toes are equal, the two sides two pads are equal, and the planter pad has two distinct lobes at the hind end. There is an obvious cross shape within the area, below the top toes and to the sizes and either side of the two lower toe pads. Most dog prints will show this, especially the pugs that are not open, which means the toes are outstretched. A straight cross with unbent lines cannot be drawn within any cat footprint because they have different sized tow pads and they are at different levels. Only a cross with bent arms can be drawn within a cat print and then it is usually on the hindfoot, the one that is most likely to be confused with the print of a canine. Fox prints and wolves are very similar in shape. These prints are from the more natural type of dogs, such as collies and retrievers. The claw marks are also distinctive in this case; small dots are placed on the ground where the claws have pricked the soil or snow. Most dogs have claws that are long and penetrate the ground, especially mud or sand.

Lonesome Cubs

In 2006, I went out on my usual roadkill drive to Salisbury, then to Wimborne. I did this at least once a week if I felt well enough. For the whole of my adult life, nearly all of the meat I have eaten has originated from wild animals killed on the roads. I also have always done a bit of taxidermy, so finding dead animals was important to me for two very good reasons as well as others.

It was late April and it was sunny and rather warm. I had two dead pheasants in the car, and I was happy with them. I drove through the corner of Cranborne chase and parked on a hilltop. I watched several buzzards displayed on the thermals. I was watching them through my binoculars. I had a sandwich in one hand and bins in the other. It was about midday. I scanned the fields below, looking at the new growing crops and something caught my eye in a field below a large area of woodland.

At first, I thought that they were cock pheasants fighting, as this was very common at this time of year. I then assumed them to be black pheasants. These were common in some areas, they are actually dark bluish-green, but from a distance in the shade, they do look black. I just saw dark shapes with long tails running around in the distance. I then realised that they were far too big to be pheasants.

The bins went to the eyes; the sandwich fell to the floor. What the hell is that? Two animals were darting around to and fro on the field of newly sprouting wheat. They were about half a mile away. I could plainly see four legs, not two. They seemed to be working together. They were jet black and very long in the body.

At first, I thought that they must be dogs and an owner would appear very soon. No, there was no owner. No dogs.... Just large cats, chasing what I presumed to be leverets. This field was good hare breeding and due to the odd turns they were making, I can only assume that they were trying to catch the baby hares, which can run very fast soon after birth. They were certainly cats but rather small, about the size of very big domestic cats. They had long, thick tails. They may have been cubs, but if so, then their mother was nowhere to be seen. They did not look big enough to be large cats or big cats in the sense of leopard or puma, but they could have been hybrids between a puma, domestic cats or some other larger species. They behaved in a similar manner to long dogs coursing hares but were clearly cats, not dogs. In the same area, there had been many reports of huge black cats and some light brown cats. There had also been several reports of lynx within the Cranborne chase areas. I had found scats and footprints of possible lynx, just a few miles away and just a few hundred yards away from where I was parked, I found a road-killed body of a cat that fit all the hallmarks of being a wild cat. I will mention more about that later.

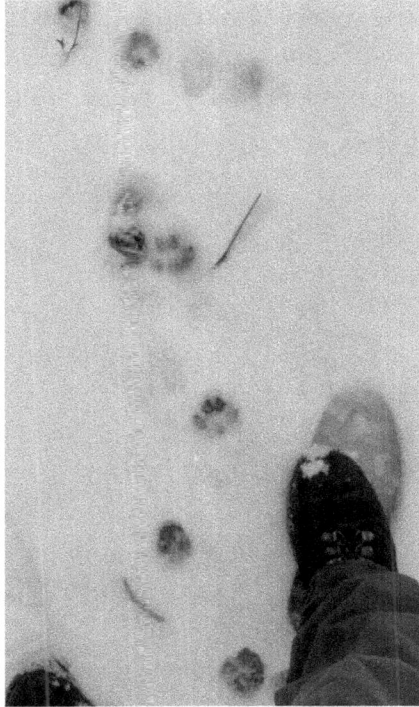

These are the tracks of a Dorset wildcat (Felis sylvestris) in Cranborne chase near to where I found the road-killed wildcat kitten in 2006.

Two Species Within An Hour

In around 2003, large cat enthusiast Danny Nineham visited Dorset and we went out searching for cat evidence. In Wareham forest, we found the remains of a sika deer that had fallen from a tree during gale force winds. We found typical big cat evidence within the large woods, including scrapes, footprints and spraying areas. The same night we went on a drive and saw a lynx casually walking down a rural driveway to someone's garden. A few hours later, a huge black leopard crossed a road near Cranborne, causing pheasants and crows to make alarm calls as it ran through the woods at our approach. It was that memorable night that I also sniffed out a decomposing deer from the car window and stopped, only to find a roe with coalesced antlers! I took the specimen and it is now mounted on my living room wall. It may have had nothing to do with large cats, as it was just a roadkill without any evidence of predation from cats, but it was something even rarer!

Possible lynx footprint from Longleat estate, Wiltshire, in snow.

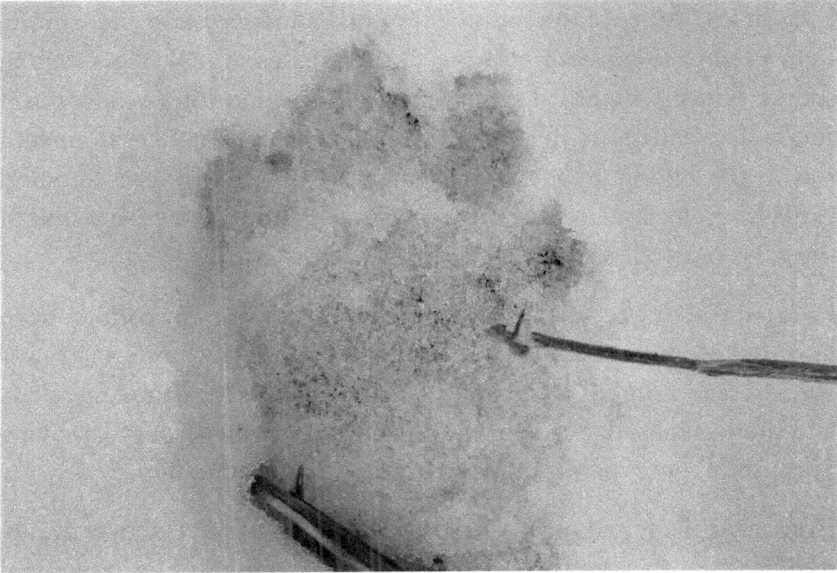

The hindfoot of the lynx at Longleat.

The Heathland Parent

In 2006, I decided to look for spiders. The summer before, I had found a species of Dolomedes spider that looked like plantarius. Dolomedes fimbriatus was the more common of the so-called raft spiders and the fen raft spider (Dolomedes plantarius) was only supposed to exist in Norfolk fenland and a few other sites in southeastern England. However, I had found a few that seemed to look like the rarer type, so I decided to visit this particular area of wet heath in the hope of finding another specimen.

It was a very warm day at the end of September. I waded through the high tussocks of purple moor grass. I was on a deer path but could not see where exactly I was putting my feet. I was heading toward an edge where a small stream ran through some dense willows and gorse with rushes. As I could not see where I was going, I walked slowly so as to not accidentally hit a bog hole. There had been recent rain and I was being careful.

I was at first aware of the sound after I had paused to stop a few times. I am used to the breeze blowing in the grasses; it makes a swishing sound, but just slightly. I was aware of this sound for a while, but it suddenly occurred to me that there was no breeze at all. I stopped. I was in the middle of the bog area and approximately one hundred yards from the stream. Yes, there it was again, the distinctive sound of the grass moving, but why?

I started walking again and realised that the sound was coming from in front of me. That's odd. I stopped again, but nothing. I was surprised and slightly suspicious. I carried on this time looking ahead and then I realised. Something was moving up ahead of me, moving purposefully in sync with me. An animal was in front of me.

How strange, I had been walking in this grass for about twenty minutes, slowly pottering about and there seemed to be an animal walking in front of me! I could see the grass now moving, only the tops. The grass was about just above knee height, some places slightly higher. The animal was shorter than three feet in height. My thought then was why an animal walking in was in front of me and not aware of my presence. I was not being exactly quiet, on the contrary, my feet were making slight squelching noises at times and the grass was swishing. Before in the past, I had managed to fool deer and other animals, especially badgers, by making a racket just like they did and they would assume that I was just one of them, but I could not see any deer, besides it would have to have been a very small roe if it was a deer. Maybe it was a fox. Yes, that was it, a fox that was going deaf; it happens. It seemed very odd, though, that it was keeping the same distance as I moved on. It suddenly dawned on me that maybe it was doing exactly that on purpose. But why would an animal do that? I then had a bit of deja vu. Had this kind of thing happened before, or something similar? Gosh, yes, the puma when I was a kid. I wonder if it is a cat. I still could not make out the animal, but I think I could make out ears that were held back, maybe of a greyish colour. That was it. It was a cat and I then knew precisely what it was up to. It was a cat trying to lure me away from her cubs. Well, I'll be damned.

The cool, calmness of the animal was amazing. The deliberate sussing out of me, then the controlled, deliberate thinking, was awesome. It worked. I had moved off towards the animal, so it clearly wanted me to know of its presence but not overwhelmingly. It was subtle but clear. *Shit!* I thought. I abandoned my spider hunt and turned on a halfpenny. I did not want to disturb it anymore. I must get out of here. Poor thing must be terrified. I had entered a quiet remote area where no humans walk just to look for a damn spider and now I have disturbed a she-cat and cubs.

I had walked about twenty yards and then realised that I had been walking on a muddy path, concealed under the bent-over grass. I parted the grass as I walked and stared down into the brown mud. Several paces went by before I could clearly make out the pug marks of a cat. Small though she was, and certainly puma, the hind right side foot, then a forefoot, just a bit bigger than a large badger's footprint. I then hurried out back onto the path.

It took just ten minutes about. When I was on the dry heath path, I looked back in the direction of where I had come and tried to suss everything out. Where could the cubs be? There were many places. There were many little dry islands amidst the bog and moor grass with gorse bushes on them. Maybe that is where she left them certainly, they were the safest place and if I were her, that is where I would put them. Just then, as I started to walk back towards the car, two sika stags ran very fast across a field in the distance. It would have been about two or three hundred yards in front of the cat. They leapt away and into the far woodland. Wow. What a day. I must get back here and look for those cubs. I must come in the morning, first thing.

It was a cloudy morning, and I wondered why I bothered. I had got up too early. I had been thinking about it all night. I had slept but with an overactive mind and the imprinting of the fact that I must get up early meant that I awoke at four o clock. I had to get up; otherwise, if I went back to sleep, I would not get up again, so I forced myself to get out of bed. My body was aching and I went dizzy.

Living with fibromyalgia was difficult. It was a curse. The only thing that could override the feeling of fatigue was an adrenaline buzz. I was out and near the spot at five am. I parked the car under lots of oak trees and just sat there in the car for what seemed like ages, watching many greater horseshoe bats flitting to and fro. Wow, this was good if nothing else was going to be. I had the windows down, listening. All was quiet. I could hear the distant rumble of a ship in the harbour. I had decided to take my big camera lens, six hundred mm and with a times two converter, on an old Pentax body. That should do it and with a heavy tripod, within an hour, I was at the area.

I crept along with my combat colour clothes, tripod across my shoulders. I reached an area in from of the patch of bog and moor grass. I was just off the sandy heath path and decided to put the camera up by an old holly on a dry patch of heath overlooking the area where the encounter took place. It was just starting to get light. I was cold, but I stuck to it. I did not have a hope in hell of seeing the cat or cubs. I did not believe it for one moment. She had probably already moved them much further away.

The mist rolled around and I could see nothing. There would not have been much of a view anyhow as the whole area was flat in front of me. Maybe I should have gone on the low hill behind me. No, there were birches obscuring the view of the moor grass beds. I shivered, hunched up against the holly.

As dawn approached, I was going to go back home. I could see a herd of sika in the distance about a thousand yards away and another smaller herd near to me, actually about two hundred yards away. They were not feeding but lying down, one was standing. I swept my binoculars around to notice another sika lying even closer to me but away from the herd. I casually wondered why it was not with the herd.

I sat there for nearly an hour. The furthest herd had got up and slowly moved around, some grazed, others milled around. The nearest herd actually got closer to me whilst feeding on the heather flowers in a dry area. I did notice that the deer that was on its own was still on its own. I sort of subconsciously wondered why but did not think too much about it, just a fleeting thought. It was the size of a yearling just led down.

The mist was clearing but still thick around the damp areas. The deer herds were now out of view, swallowed up by the mist. Nothing here, I shall go home. A tawny owl hooted and I was just about to take that as a sign to go when out of nowhere, like a bullet, an animal just leapt from behind a gorse bush and headed for the lone lying deer. It all happened so fast I could hardly take it in, so fast it came from out of the blue and maybe five or six long leaps toward the deer. It reached the deer and the deer just leapt vertically into the air. Straight up about five feet or more, and from that lying position, and as it did, the other animal passed underneath, turned and then both animals leapt back the way the first one had come.

Shit. The deer was not a deer. It was a cat, and the other animal was a cat. They both were cats. As the lying down animal leapt high into the air, I could plainly see its muscular body and long thick tail. It was grey in colour but so athletic, that it turned in mid-air. By the time the assaulting animal had got to the other, only its back was visible as the vegetation was obscuring it. Both animals were the same size, colour and quickness. I was watching two puma cubs playing. The drama was about three seconds long. I had been watching a puma for an hour, thinking that it was a damn deer!

It was only later that I kicked myself over the lost chances. Not to mention the camera, idol redundant on its tripod. Wow, and how stupid of me. That was it. No more. A quick blast and that's all. I stayed for ages, but as it got light, there was nothing. I walked back to the car with mixed feelings. On one hand, I had managed to see the cubs, but on the other, I had not even a photo.

I sat in the car pondering over the situation. I kept on asking myself how I managed to mistake a puma for a deer. Also, the puma must have seen me. I was there for over an hour in plain visibility, mind you, it was looking away from me, and what about the second cat that ran at the flirt. Did that also know of my presence or not? I suspect that I was being very quiet, and maybe it did not see me. It was, after all, behind the gorse bushes. I wonder if the parent saw me. She couldn't exactly rush in front and try to guide me away. It has now happened twice. A mother puma was being very cautious and collective in the face of danger, risking all for her cubs. I wonder how many pumas had done that in America only to be shot and their cubs starved to death? The cubs, on the other hand, were large, if I mistook the first for a sika hind, then it was very large. They must have been semi dependant, nearing their departure. Pumas stay for an average of one and a half years with the parent leopards, on the other hand, may do so for two years.

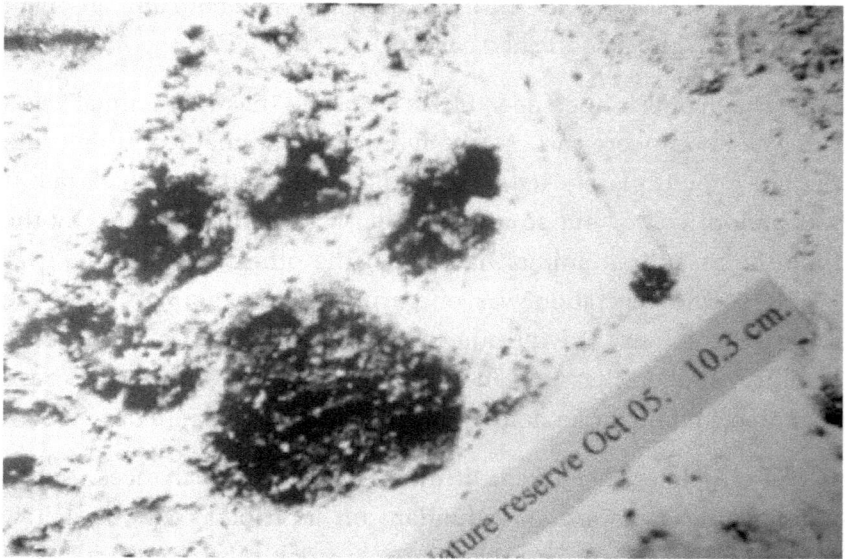

A puma print from Arne, Poole harbour.

Poole harbour plays host to several reed cats, otherwise known as swamp or jungle cats. (Felis chaus). They leave scats and footprints around the reedbeds. The footprint is six and a half cm wide

The Firing Range Cubs

I drove out with a fellow big cat enthusiast David Mitchell and his son. We went to a military area where there was a good view for miles across heath and moorland. Included was a large quarry. I had seen moving animals around the quarry before and I just knew that it was a prime area to watch. Dave was eager to see a big cat and he even acquired two trigger cameras for us to use in our quest to capture one of these animals on film. I had never thought that such a task would be so difficult, near on impossible. These were only animals, after all, weren't they? At times I was led to think that maybe these animals were, after all, ghosts of some kind. How was it that we were getting photos of all other animals but the cats? I got Badgers, foxes, rabbits, deer, mice, buzzards, ravens, you name it, we had it; all except the cat. I cannot tell you how bewildering this was.

There were logical answers to this question, though. We had had the cameras up for over a year and nothing, although no picture seemed to show a large furry outline of a head, but it could have been anything and it was overexposed. We were losing faith. Dave was not expecting to see anything, nor was I.

It was early April, quite mild, with no wind. We planned to drive around the roads with a searchlight mounted on the roof of the landrover and a camera running to secure images of anything that moved. I decided to stop and look over this particular quarry as there had been several sightings along this road over the last few years. It was on the edge of a no go area for the public. Huge herds of deer roamed without much bother. No dogs to chase them or loud kids on off-road motorbikes. The deer were relaxed. Or were they?

We stood with binoculars in hand, scanning the white glow of the chalky soil in the quarry ledges. It was getting dark, the light was fading. The sky was clear, though. The quarry was large and sprawled over several large ledges, some with mature trees still standing on what looked like islands. An unworked area lay above and to the right-hand side of the overall area. I noticed a deer running very fast across the bottom-most part, a ledge of a white diamond-shaped area of lightly sloping chalk. That was odd, the deer, a roe, as I just had time to catch it in my binoculars. Deer do not do that unless spooked.

I said to Dave, "check out the diamond triangle, a deer just pegged it across there."

Dave seemed to suggest that he had his eye on that particular area. The quarry was a mile away. I knew that the deer was spooked, yet there were no people to be seen. There were no dogs on the loose running riot. It seemed very quiet. I scanned the top again, and then the worked areas, the high cliffs, the fields behind and then back down to the white triangle near the bottom.

There. What is that? I strained to focus in the fading light on a black object, motionless in an area that I was sure was empty a few moments ago. Yes, definitely an animal, then it moved. Wow, shock.... A distinctive shape of a large cat was walking up the slope, then another behind it; exactly the same. Two animals, very dark in colour, walking and pausing, sitting down on their haunches twice before walking up the slope. Their dark colour in comparison to the dirty white of the substrate was very clear. They were cats and they had very thick tails. They were very muscular but looked quite plump, not skinny at all, not even slim but bulky. Gosh, they must have been huge because they seemed to be the size of the deer that legged it across the white patch, at least their body size was the same, if not the legs, and they did seem to be lower to the ground.

I frantically urged Dave to see them, but it would seem that the white diamond-shaped area was not the same one as I was looking at or had been for the last half an hour. I had watched these animals for about five minutes and was mentioning it all the time to Dave, but he could not see them. It only occurred after they had gone out of sight that he was looking in the wrong area. I think that I was more ashamed than he was. I just could not understand why he was looking in the wrong place. We were bitterly disappointed.

Afterwards, I was just sad all the way home because I, on the one hand, really wanted him and his son to see a large cat; on the other hand, I maybe felt a bit guilty in the fact that I did not describe the area properly when I had seen the deer run across. Unfortunately, there was another diamond-shaped area but at the very top of the quarry and not the bottom. I am sure that I gave good enough directions and said that it was in a straight line from the field we were looking over. Well, never mind. I just hope that he believed me.

What I saw were definitely two very large cats, both of equal size, so one would suggest that they were cubs. They had the body size of an adult roe deer, with long thick tails more like pumas than leopards. They would have been the size of large retriever dogs with thick short legs. If they were cubs, they would have been at least in the second year, possibly had already left their mother and maybe had been hunting together, but I think not. Just weeks beforehand, I had found two deer carcasses within a mile of each other about a mile from this area. The sika hinds had been totally stripped of flesh, and the area around the body had been flattened by more than one body. A small scat of deer hair was left a meter away and the carcass showed all the hallmarks of being eaten by a large cat, or two or even three. Oddly enough, another carcass, again of a sika hind, was found about three miles away; again, it looked as though more than one animal had fed from it. Surely this could not have been the same family?

The Airport Cat

Another sighting I had was of a beige coloured animal, possibly a puma, near Bournemouth airport; well, it was just behind the long runway on heathland. There had been many reports of a big black cat around the site and people had seen it around the industrial estate within the airport perimeter. I had also picked up spoor from a local wildlife trust reserve alongside the eastern edge. A story goes that a few years ago, a large airliner was prevented from landing as a big black cat was sitting in the middle of the runway! I heard this from one of the airport control crew. There is thick heath and woodland surrounding the airport on approx half its side. I was searching for sand lizards in a small nature reserve when something caught my attention. A wren was giving alarm calls from a bramble bush, so I looked in its direction to see two hind legs disappear behind some small conifers. I immediately realized that the legs were short and thick and of a uniform tan colour. They were crouched as if the animal was creeping along. They paused for a few seconds before they completely disappeared out of view. The animal was careful not to run away but hide and stay within the area immediately in front of it. I realized that the animal was a large cat and possibly a puma by its large size. I only saw the legs through a gap under the bush. I do not think that the cat realized that I could see it as it tried manoeuvring around in such a way as to keep it hidden from me, but it was only partly so. Had there not been a gap in the bramble bush, then I would not have seen it; however, it just goes to show how some cats will not move far but just hide around the immediate vicinity.

The Molehill Leopard

During 2008, I drove out to one of my best study areas at dusk, and slowly ambling through the lanes, I entered a hot area by military land. This was an area with firing ranges, heathland and oak forest. There were grass fields on both sides of the road with rather high banks with hedges. It was to my left-hand side that I noticed the silhouetted shape of a panther looking down at a molehill just the other side of the flimsy hedge. I slammed on the brakes and reversed, but the cat legged it very fast to the field edge. It was jet black although the sun set was behind it. The typical pantherine head was obvious and just staring down at the ground at the molehill which was by its feet. It reminded me of when a dog looks at something strange and moves its head to one side with curiosity. It would have made a fantastic photograph, but of course, there was no chance as it all happened in just about seven seconds.

The Hartland Puma

During the same year, I was driving through Purbeck heathland slowly and it was well after dark. I was on my way home after watching an area where rabbits, deer and badgers foraged in a large area overlooked by a hill. I enjoyed watching these animals and listening to cuckoos and nightjars just after the sun had set.

I was hoping to glimpse a large cat in this area but was not so fortunate, so I waited till it was too dark to see even with my binoculars and decided to slowly go back home via some back roads through the heathland. The sky was a lovely orange, and mists started to envelop the valleys.

When I approached an area of heath with a slope to my right-hand side, the car headlamps picked out two bright yellow, orangey white eyeshine. The animal was walking down the slope towards the road. I slowed down, not knowing what it was at first but realizing the eyeshine was not a fox, badger, or deer thought that it could be a large cat.

As we both neared each other, it was clear as to what it was. It was a puma, dark grey-brown in colour. It was slinking along a barbed-wire fence boundary by some ponds. There were large stands of bracken and gorse nearby, but the cat was in the open as no vegetation was growing on the slope and deer and horses had grazed the grass and heather very short. I had a full side view of it. It was darker on the back and the tail was slightly darker as if a continuation of the broad darker dorsal stripe. The tail was very thick and darkening towards the bushy tip.

As I got even nearer, the cat seemed to walk in slow motion, as if it was deciding whether or not to continue or turn and run. I made eye contact with it as it crept on its belly down the short grassed slope. It was looking at me! Right at my face, at my eyes. It saw me, not my car. As I pressed the switch to lower the window, the animal turned on a sixpence and loped off up the hill very fast. As it did, I just caught sight of the body within the light from my headlamps and realized that it was a large cat. Its rope-like tail was held to the side as it ran and it hid in a large area of bracken. It was a big puma, I would suggest male due to the large head in relation to body size. It was very thick set and healthy-looking. There wasn't much point waiting around to see if it returned. It was heading towards the areas where the sika deer graze and much water about. In this area of the heath, many bones can be found, remains of their kills, and scats.

Hartland Moor in winter, an area where all three large cat species leave their mark. There may be fewer people about outside the summer months and large cats can be themselves to some extent and need not keep such a low profile. There have been more cat sightings here than anywhere else in the county. The calls and screams of puma type cats have been heard for decades.

The Motorway Cat

I decided to visit one of Britain's largest antique fairs at R.A.F Swinderbury in Lincolnshire one year with a friend of mine. It was a long drive but a lovely sunny day. Near Silverstone, we took a slight detour from the motorway and as we did, we reached a large intersection of motorway junctions and flyovers. There was a bowl in the land surrounded by roads and a farm was nestled within this round valley like land hemmed in by busy roads and traffic.

I suddenly saw an animal sitting bolt upright in the middle of one of the slopes surrounding the bowl area near to the road we were on. My eyes locked on it and I was suddenly shocked as I realised that it was a huge big cat. I shouted to my friend Mike about what I had just seen, but we could not reverse and have a closer look as we were on a busy one way stretch of road. I did get a clear view of it, though. The animal looked frustrated, lost, and confused and did not know what to do. It had obviously got caught out whilst travelling, perhaps looking for a territory of its own or exploring, maybe a cub just left its parents. The animal was huge, almost the size of a Great Dane dog or at least a German Shepherd! It was looking right to left and back again as if contemplating what to do! I felt sorry for it. I hoped that it would remain there until the early hours of the next day, when it could find its way out with less risk of being hit. It had a long wait as the time was only 8.30 am, so I assumed that it had been travelling during the night or at least the early morning. The colour was mid chocolate brown, but not particularly like a puma. I saw no other colour on its face, but I did notice a long head more like a leopard than a puma. Its tail was curled around the base, so it was out of sight, even though the grass it was sat upon was short. It looked very sleek but muscley!

Again I was confused as to what exact species it was. One would assume that when one sees these large cats clearly that one could identify the species, but it isn't as easy as that, oddly enough! And I don't think that it comes down to just lack of experience. The cat looked like a leopard, but chocolate or liver brown like a Labrador dog. It certainly was lighter on the undersides and belly and chin. I saw no spots, but often, when one views a leopard from a distance, the spots or rosettes cannot be seen and the animal may look uniform in colour; I'm not sure that this was the case here.

The Dartmoor Cats

The west of England may be the most highly populated area for large cats and certainly has more sightings or other reports than any other area for longer periods of time. Dartmoor and Exmoor are both quite near each other straddling Devon and Somerset. I had been fortunate enough to be invited but wildlife film-makers to advise and search for large cats in several areas of Dartmoor. I found much evidence in the form of scats and footprints and saw a caracal along with fellow researcher Chris White. We were asleep in a land rover within a steep Devon bank within a remote area of forest. We had been tipped off that a pair of elderly caracal had been released. I found a footprint, another of the team found a whisker and we all found bones of sheep and deer. Many scats adorned the private forestry tracks. The caracals bred and produced one cub according to a military-based person who had computer software linked to satellite technology and had seen them on such equipment. Six people were in the car when only Chris and I woke up to see the caracal leap over ten feet across the ditch over the roof! It happened so quickly, as if we were both dreaming, but we knew that it was real.

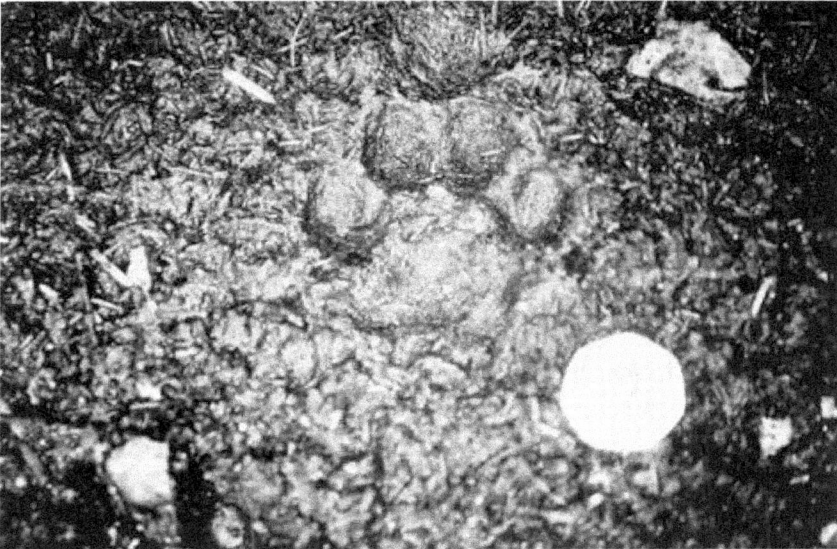

I visited Dartmoor national park several times to look for evidence along with several other people and we were told that there were caracals living in an area. I found this footprint that may be from a caracal in the area where they were thought to be. A whisker was found by somebody else also, and many scats.

The same year, I saw a large black cat sat on a hillside just outside Widecombe in the moor, seen only by myself whilst the team was out shopping, ready for our next adventure. We turned the car around to where I said it had been, but it had gone.

Dartmoor has had more than its fair share of large cat sightings for decades and I would have been very surprised had I not found any evidence, sightings continue and today, there are several investigators around the area getting evidence and following up on peoples' reports.

The Eyeshine Of The Beast

On many occasions, I have seen bright eyeshine of animals that could have been large cats. Most mammals' eyes will shine when light is directed at them either by a torch or by vehicle headlamps. With experience, one can make out many species as the colour and intensity of the reflection varies.

All our common mammals reflect as white from certain angles, but when the head is moved to the side, one can see different hues such as green, yellow, blue, amber or pink.

Nocturnal animals have good night vision and their eyes let in more light that reflects when bright light hits the retina. The better the night vision eyesight of the mammal, the greater the reflected eyeshine, and cats, having the best night vision of our mammals, reflect greatly. Nocturnal mammals have rod cells within the retina, and diurnal mammals that can see more colour, have cone cells within theirs. The tapetum reflects the light.

When I saw a puma for the first time, I was amazed at the brightness of the eyeshine and compared it to other nocturnal mammals and realised that the shine was greater and far more obvious. The puma had intense yellow eyeshine when it looked face on to me, without green or amber as one sees for deer, badgers and foxes. I have seen this bright eyeshine many times when scouting a bright torch across areas of the countryside such as fields. (One must also watch the movements of the animal and by the way it moves its head or body about, one could see if it is a cat or other mammal. Most cat species show up with yellow or orange-yellow eyeshine and red from certain angles. Many people report cat eyeshine as being green, but I haven't seen this myself. I have seen photographs of bobcats and wild cats showing emerald blue-green.

With the old stories of black dogs haunting the woods and fields in olden times, it was noted that their eyes would glow like fire or embers. This reflects the large cat's typical colour of eyeshine in certain types of light, and it is one of the many traits that lead many other people and me to assume that the stories of black dogs at night are most likely to be large cats, and most likely the leopard. I will touch on that subject later.

Cats will blink much more than other mammals as they have such good night vision and it is uncomfortable for them to stare at a bright light at night, so they will turn their head to the side often and blink lots. Cats have enlarged rod cells within the retina adept at letting in more light on the subject, thus allowing them to see in darkness. But having evolved this, something must give that is not so important, and that is colour vision.

Birds and reptiles, and fish need colour vision to see leaves, fruit, and seeds and dangerously warn insects and snakes. So the cat has done away with good colour receptors and replaced them not entirely by light-sensitive cells. Animals that have a good colour ratio have cone cells within the retina. An animal that drops to the ground is a dead giveaway as most other mammals do not do this (but roe deer, badgers and foxes are capable of doing it when they think that their life is in danger). Also, the eyeshine of large cats can be seen from a much greater distance than other mammals. Some people gauge the distance between eyes to assess whether or not it was a cat, but that is meaningless as most large cats' eyes are the same distance apart as a fox, badger or deer, the latter being slighter further apart, especially in large mature bucks or stags. One thing that is true is that the eyes of predators are forward-facing and so both eyes will be seen clearly even when the head is slightly turned to the side. When deer turn their heads slightly, one eyeshine is lost due to the fact that they are more on the side of the head rather than at the front.

I have seen a badger drop on its belly to hide from my approaching car, and I have seen roe deer drop to the ground and hide in a predatory like way, but only because it was trapped by its foot in the lowest strand of a barb wire fence.

When the London cryptozoology group visited me and my research areas in 2013, a night drive in prime habitat shoed such bright yellow-orange eyeshine, of an animal which was walking along the edge of woodland in a hilltop field where deer always graze at night. When the car slowed down, the animal sped across the field like lightning, across the road in front of us and through into the next field and along the fence line. It ran very fast and the eyeshine could be followed even though the animal was side on all the time. I knew that this animal could only be a large cat. The behaviour it displayed was not the same one would expect from a deer or a fox. All the passengers in the car saw it and agreed that it was spectacular. The animal made a decision to bolt as soon as my car's headlights hit it, and it ran for a great distance and extremely fast. The whole behaviour of the animal suggested something far more intelligent and shy than the usual wildlife.

When we held the first Dorset Big Cat conference near Wareham in 2013, a short walk from our campsite conjured up the same eyeshine of an animal quite a distance away, walking swiftly along the bottom of a hill. Several people also saw this and agreed that it was more likely a large cat.

On at least six occasions, whilst driving through cat areas or my study areas at night, I have seen the same eyeshine reflected and the animals have either been waiting in woodland or seen up ahead on the road in the distance, out of range of visually being able to see the animal within the headlamps, but the eyeshine penetrates.

The animals that have the most sensitive eyes are no doubt dazzled by the brightness and so will blink more often, even closing their eyes for long periods of time or turning its head away from the source. The cat is the most likely mammal to do this.

The Cotswold Puma

Another brief sighting was with friend and big cat enthusiast Rick Minter. We were driving around the hilly areas between Gloucester and Stroud (as we often do when I visit him) and we had Chris Johnston with us in the back of the car. He is also a renowned expert in the field of big cats.

We were driving through an area of beech woodland along the base of a hill when I spotted a large cat sitting in somebody's long gravel driveway on the edge of woodland. I shouted stop and asked rick to reverse a few meters. On doing so, both rick and I viewed a puma-like cat that was sitting bolt upright on the edge of the driveway. It was looking towards the house and when the car stopped, it looked round before leaping into the woodland. Unfortunately, Chris did not see it as he was sitting too far back and we only had a small gap in the vegetation to see it, and it went so quickly when it knew that it was being watched. It looked like a young puma, but we could not be sure. It was slim and lanky and light tawny brown with a clear lighter underbelly. The head was rather small in relation to body size but not maybe as much as expected from a female puma. I suggest a young male. The tail was long and rope-like, not particularly thick, but not thin either. I felt that the cat was in one of those careless playful moods that cats go into, especially juveniles. The sighting lasted just a few seconds. We did, however get out and walked into the woodlands only to hear a trail of bird alarm calls going up the hill. Robins, great tits and crows and magpies fired off for quite a way up the hillside. There were some small dark scats dotted about a grassy area nearby and they could have been cat but were small.

The area had been riddled with sightings of large brown and black cats. I often visited Rick and his family about twice a year or more and we would go out and look for evidence or speak to people who have had sightings and to put out game cameras. This part of Gloucestershire is rich in large cat evidence and with many enthusiastic people on the case. On another occasion, when Rick was driving nearer to home one dusk evening, I glimpsed what looked like a large light brown puma fast trotting across a small paddock adjacent to the road, but when we reversed to get a closer view, it had already disappeared. It was in front of a farmhouse with livestock. There were no other people around the property and it was quiet. It may have sped up as it noticed us slowing down. Any suspicious activity from the norm will cause a cat to change its behaviour. Cats know the usual continuous movements of cars and many cats know that people are in them and so when a car stops, people often get out. If vehicles stop, doors opened and slammed, or peoples' voices suddenly erupt, then the cats will be off. It is a bit like buzzards and cars. One can see many buzzards sat on telegraph poles along the roadside and the cars go right underneath them or beside them and they don't bat an eyelid, yet if the car slows down or stops, then the buzzard flies away.

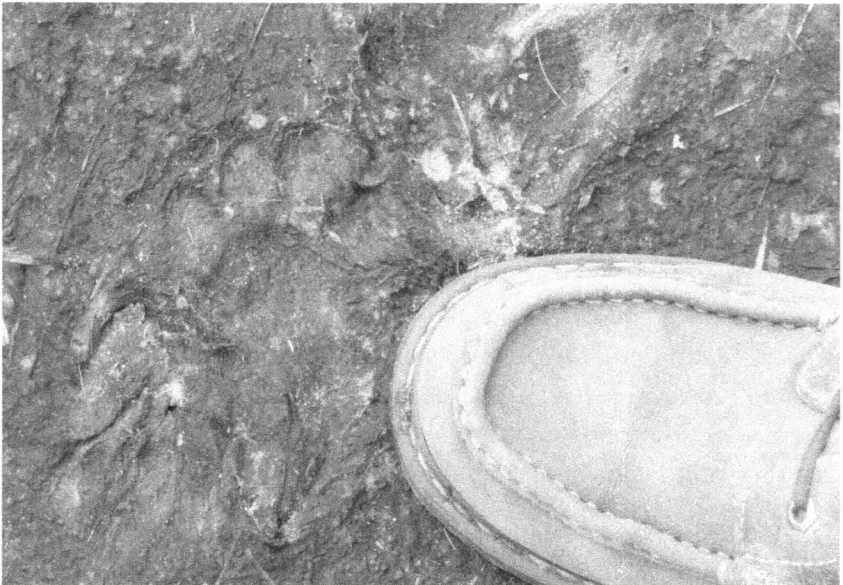

This may be the print of a leopard cub, as it lies alongside the spoor of a much larger cat.

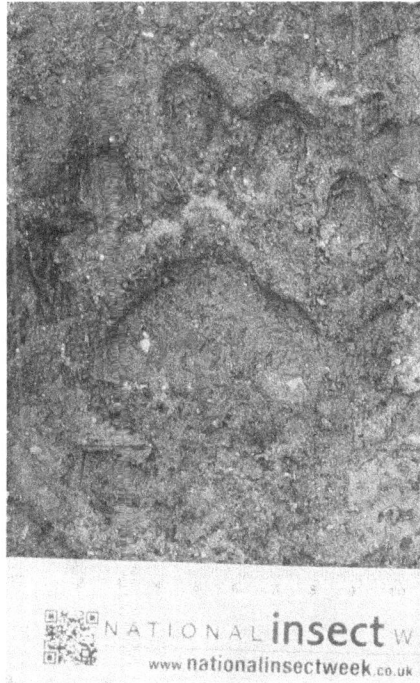

This classic large cat print could be either puma or leopard, perhaps more likely the former. I found it within an area where both species had been seen within the same week, but it is most likely to be puma going on the angle of the hind feet pads. There were two sightings within a week of a huge jet black panther-like cat. This male cat is within the airport and heathland areas and seems to have a territory between Christchurch and Ringwood. It is even active during the day patrolling its rather small territory, approximately nine miles in length and four miles in width, hugging linear contours. It also crosses the dual carriageway in its southern section but perhaps not the even busier northern section, and the river Avon is perhaps one border, and the built-up areas on the edge of Bournemouth and Christchurch, at another end. If this spoor is the black cat, then we have black pumas, or leopards have very similar feet to a puma sometimes!

Just two hundred yards away on a parallel track, a cat made a string of tracks in the mud alongside the dual carriageway on heathland. It paced back and forth actively, perhaps following a female in oestrus or trying to avoid another large cat. All prints may have been from the same individual, but there was a slight size difference and the sandy mud prints looked more like a puma, but the woodland print looked more like a leopard. Two big male cats of different species will try to avoid each other but confrontations will inevitably occur. I got the impression that the puma was the one breaching territorial rules and the big male leopard was uneasy about it. Its footprints covered a large area zigzagging back and forth across the dual carriageway. The cat rarely walked at a leisurely pace but in a hurried way, not showing direct register for two-thirds of the time. Scrapes were found at the woodland site in usual places by the leopard and scat was deposited in the mud area alongside the dual carriageway. Forefoot.

*The puma mud tracks hindfoot showing three-lobed back pad
and slipping.*

The left forefoot of the female Avon leopard. Both she and the main male of the area often drank from a heathland pond. As it dried up during the summer, the cats would walk out, leaving behind green footprints in the algae-covered mud.

This photo shows the high angle of a hindfoot of the small Avon female leopard on sandy heathland.

The Male Avon leopard left a footprint as he sniffed a gate post.

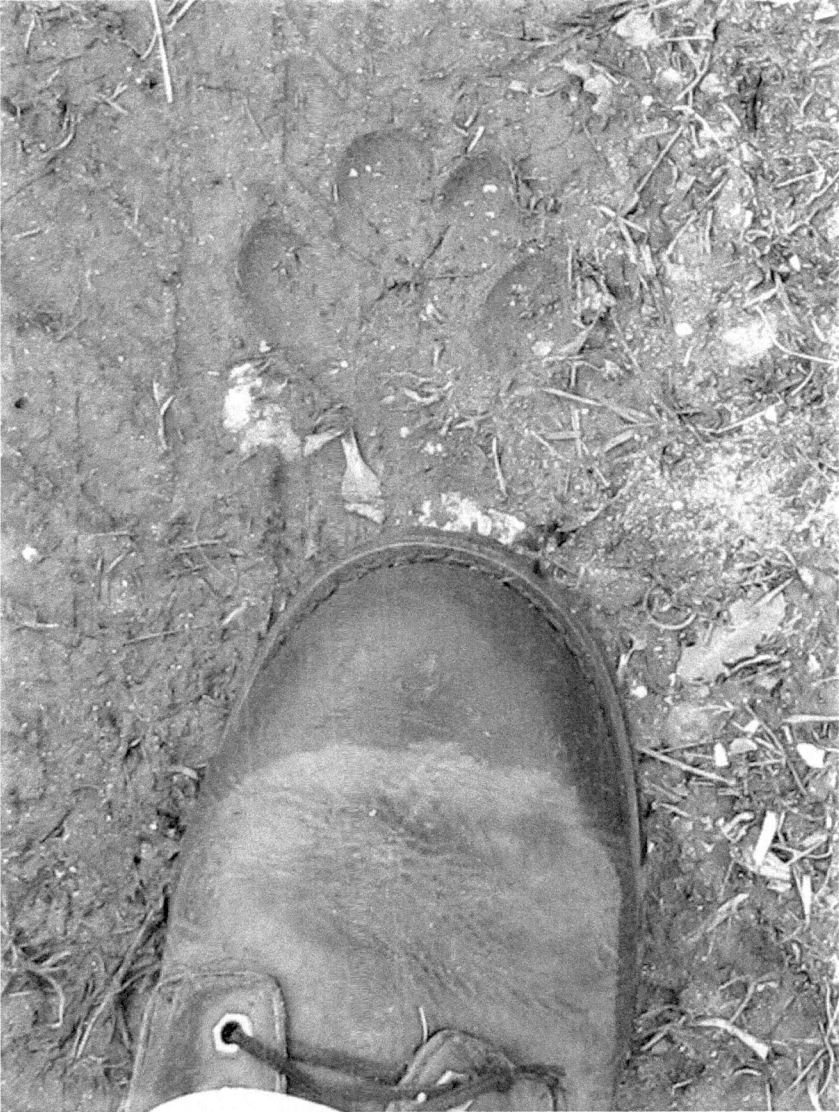

The three counties male leopard left his hindfoot mark high on the downs.

The female Avon leopard left her prints whilst trying to get one of the drying up water sources.

The male Avon leopard also left his mark in the drying up heathland ponds.

After the rain on a usually dry hilltop near the three counties' borders, the leopard has left its foot mark. There were several strong-smelling sprays nearby. Sometimes it is difficult to decide whether it be a male or female. Usually, female feet are smaller and a bit slimmer with more angled hind pads, but some male hind feet are slim also. Many times it is also almost impossible to tell what species was responsible, whether it was a small leopard or a big lynx or a puma, as sometimes only parts of the print are available to see. It is the hind pads that differ mainly with the puma having a more angular structure on the hind feet, leopards are more usually near level, and the toes are more often just almost level but not quite. They can look very much like dog prints, but the differing toe sizes, slight misalignment and large three-lobed level heal pad suggest leopard.

Large cat footprint found in Winborne.

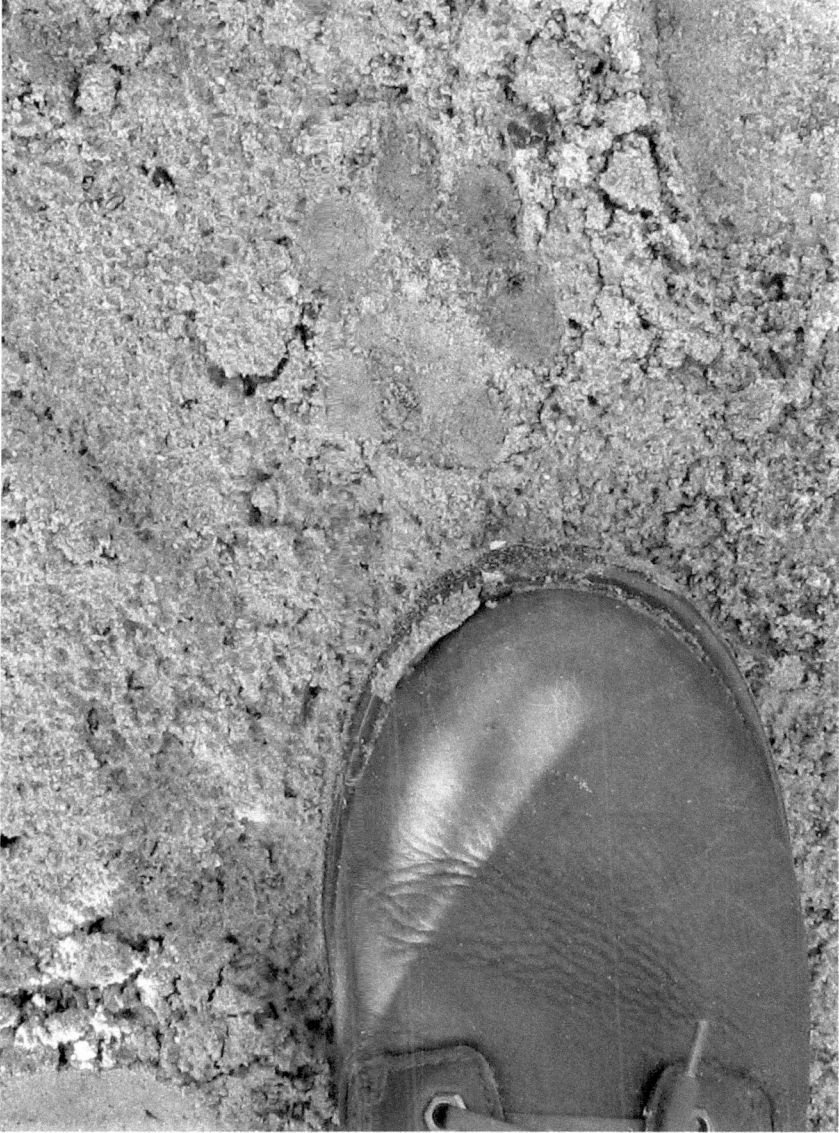

Avon female leopard hind foot print.

Avon female leopard hind foot print in dried bog.

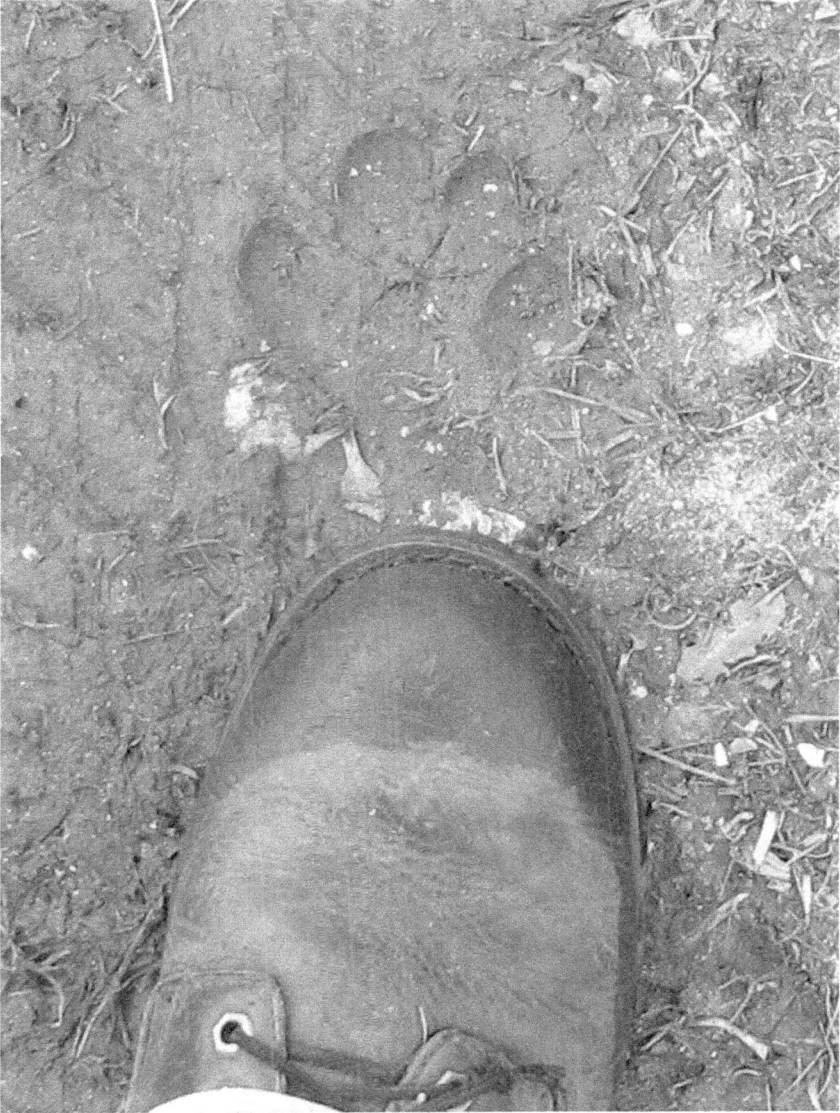

Avon male leopard foot print.

Male leopard spoor.

A captive puma shows its toe and heel arrangement.

I also spotted a jungle cat along with Rick and at least one other person, maybe Rick's son Owen? We were at Port Ham searching around as reports of a leopard-like black cat had been seen there. We were out at dusk in this reed bed and scrubby area of the river Severn within the city area. I first found a scat, then a footprint in a mound of soil and then saw the cat sitting bolt upright like a fox, just watching us at a distance of about eighty yards. It turned on a sixpence and leapt into vegetation.

The Midnight Stalker

During April 2014, I was in a car travelling between Blandford and Shillingstone after eleven PM. I was with my friend Tracey Saunders, and we had just picked up her daughter Etene from a party in Durweston.

Just as we reached the slope at Enford bottom, a steep incline on the chalk hillside, a large cat ran across the road in front of us and stood on the right side roadside verge and looked back at the travelling vehicle. I was the only person who saw the animal as the other two were in conversation and so maybe not as observant as they could have been. I saw the animal for several seconds and as it stood on the bank, I could clearly see the size, colour and general shape of the animal. It looked exactly like the grey version of a golden cat and was the same size. It looked like a giant moggy but with uniform grey-blue short fur with a tail just like the longest tail that a domestic cat could have. It may have had some darker grey spotting on the face and neck. It was certainly not a leopard, and not really much like a puma but perhaps more so than a leopard. It knew about vehicles adhering to roads as it acted in an intelligent manner like members of the crow family where they just keep out of range of being hit. As it reached the verge, it turned to face the car as we passed. It was extremely beautiful. It had huge hindquarters, very powerful in body and leg. The size was bigger than a fox.

The next day, I visited the spot and looked at the steep slope on both sides of the road. I found nothing of cat evidence but some rare plants; Herb Paris Carpeting the woodland. This area has had numerous sightings over the years and is just a mile from the field where I saw my first puma and, later, a leopard.

These personal sightings are most of my better or recordable sightings. I have had several glimpses of other large cats, but because they were not definite, I haven't included them. I have also been virtually on top of large cats and not seen them. Several times I had been crawling through thick undergrowth or silently searching for reptiles when an unknown animal took flight right next to me, only to shoot through the vegetation, often moving hundreds of metres out of sight. Birds such as great tits, carrion crows or magpies and jays have sounded alarm calls and would often follow the animal through the trees. Sometimes one can get the smell of a cat in these types of cases. The smell is a typical cat smell of musky, perfume-like ammonia, different to the musky coffee dog smell of fox.

I have often sensed something or someone watching me whilst I have been out and about, but to be honest, I do not suspect a cat to be watching me. But once or twice, after an animal has bolted in flight, then I often wonder, especially if birds give out alarm calls. They often do this with foxes, especially if the fox is acting erratic or frightened; it seems to excite crows especially.

Members of the crow family will mob a predator, and the most predatory the animal is, the greater the effort the crow will make to let others know. This is basically a selfish attempt to get others of the same species to respect the individual to gain status. Lots of social birds do this, including magpies, gulls, starlings, jackdaws, and even pheasants. If the predator is visible to the birds for a longer period of time, the birds may give up alarm calling unless it moves in a predatory way. Birds will also alight near cats even to pick at a carcass and so get used to the animal without needing to sound alarm calls. It is all usually to do with sudden or unannounced behaviour, noise or movements that excite the birds.

Carrion crows are possibly the most likely to give alarm calls when a cat is moving or stationary. They have such good eyesight that they can spot predators from a long way off. Usually, the crows will get as near to the predator as possible, but at a safe distance, often on top of a conifer or overhead cable, and point at it whilst crowing harshly. This can be several times repeated over and over again; Just because it stops does not mean that the cat has gone; it may be that the cat has just decided to lie down and go to sleep. The crow may then feel safer and not bother to alarm call. Sometimes magpies and crows do not give alarm calls, and for good reason. If a cat or group of cats are eating at a carcass, the birds will look alarmed at first and maybe initially give alarm calls, that is until it works out that it could benefit from the leftovers, and then decide it not best to call all corvids from far and wide to the small amount of food that could be available. Other animals may give alarm calls at cats and deer, especially if they are aware that a large cat is danger. Usually, this will be instinctive, especially if the cat is stalking. The loud high pitched whistle of sika deer can give away the presence of a leopard in the reed beds in England, just like the chital and sambar in India.

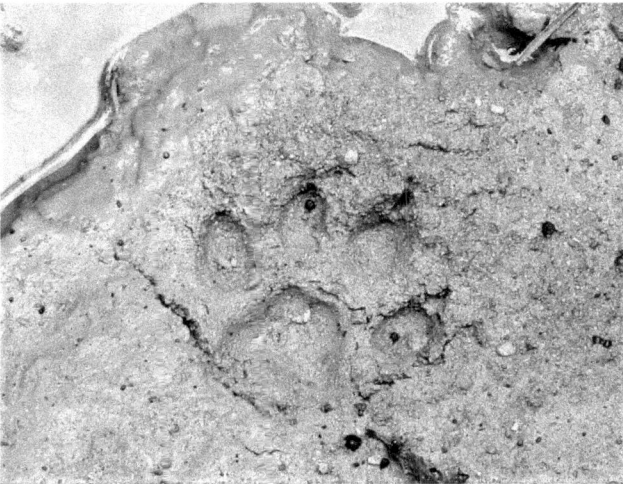

This is a photo of the footprint of a fishing cat from north East India.

Solstice Surprise

June 2018. This summer had been a wonderfully warm one and one of the warmest on record, along with the famous summer of 1976. I had a very long spell of better health, as I often do when the temperatures are above 70 degrees. I had managed to get out at least once a week doing something or another wildlife based. I hadn't been on many dusks watches, though, so decided to go out on the summer solstice and photograph the setting sun and other wildlife.

The summer solstice puma. The beige puma is hardly visible amongst the same coloured grass. 21st of June 2018. I wasn't intending to watch any large cats, but as I was within one of my best study areas, I did wonder if I would encounter anything. It was the summer solstice, and I decided to watch the setting sun on this special evening. I set up on a small down overlooking a large area of natural grassland, where a female puma often left her field evidence along a trackway. The sun was sitting on the horizon, big and orange-pink. Turtle doves cooed in the blackthorn bushes and partridges called out along with

*pheasants. Rooks clattered in their roosts and everything was
alive. The air was full of beetles and flies and moths around the
lush vegetation. I had my Canon camera on a tripod aimed at
the sun, not expecting to move it for any reason. There were
three roe deer and two hares in a field below, they started to act
abnormal; the roe darted to the middle of the field in panic and
the hares ran about with the deer. They all looked toward me,
but I was not the item of concern. There was a din coming from
the nearest patch of scrub bush and magpies and crows suddenly
started a loud racket. They made this racket for about five
minutes before I caught sight of the predator. I was at first
expecting an owl, fox or polecat but was amazed when a large,
lanky blond puma ducked from under the barbed wire fence. It
trotted down a side track before the main track, where it waited
and watched for danger before trotting along and stopping twice
to look into the deer field. The deer could not see the cat as there
was high grass in-between, yet they knew of its presence. They
could smell it and the corvids gave the game away. The puma
weaved its way in and out of the track, into the long grass and
back again in full view. It paused twice, then looked behind it. A
person was walking towards it about two hundred yards away
around a corner, I could see, yet the cat couldn't have seen her
yet disappeared into the grass. Five minutes later, a lady
appeared at the bottom of the field and sat on a burial mound to
watch the sun disappear. She sat for several minutes, all the
while, the puma was in the long grass just behind her and she
was unaware of it. The deer kept acting suspicious and actually
came forward towards the person rather than fleeing. There was
a lull in the commotion for a few more minutes as the sun set
below the horizon, and in the meantime, I was frantically trying
to manoeuvre my camera tripod to another position without
spooking anything. I managed it and took several photos with a
long-range lens, but I could not see what I was actually
photographing as there was no view on screen, so I just judged.
The lady got up and walked off and I was hoping the puma was
still there. The deer were staring into the long grass and I
assumed it was. I lost sight of it as it crept into the even higher
grasses near to where the one remaining roe was, then it re-
appeared, stalking away from me. I took another few photos
hoping to capture it. As it got darker, all went quiet as the barn
owl floated around and then a deer screamed in distress. It was
too dark to see anything, but the puma had obviously been
successful in its hunting. The photo shows the blond puma the
same colour as the dead grasses. It is facing away and the thick
curved tail casts a shadow. The roe doe is looking at it.*

I was out at one of my favourite sites, an area of great natural
biodiversity on the three counties border at Martin Down nature
reserve. I wanted to see if the stone curlews were still there also, as
the year before, I had seen a pair flying in and out of the hills to feed.

I would hear them calling as they flew in above my head or nearby. I sat behind a juniper bush with yew and scabland to my right-hand side and on my left side, I had my canon EOS camera on a sturdy tripod. Many turtle doves cooed from the hawthorn and blackthorn bushes within this natural landscape. Grey partridges trundle through the long grass and roe deer crazed and watched, forever alert. A male barn owl whisked over a hedgerow as if in an eager quest to get out and feed his hungry chicks. There was no sign of the mother owl.

There was a large field to my right side with three roe deer within it, all spaced out with one in the middle and two at different ends of the field and one hare near the middle. They had been casually grazing for the full hour, or so I had been sitting down. All of a sudden, they all barked a warning together and the two roe neared the field edge and ran into the middle alongside the other one. One roe started pronking along parallel to the woodland edge and all were staring at a certain point towards me. Two hares also ran into the middle of the field and they all looked in my direction but below me rather than at me, as they were all totally oblivious to me. I was far from them, about half a kilometre on the hillside, but I could see them well through binoculars. I wondered what on earth could have spooked them like that and wondered if, indeed, a large cat was about. I thought it more likely to be dogs being the summer solstice and people being out and about. Just then, I saw a person walking on one of the many small tracks that littered the grassland and thought, *oh dear there goes all the wildlife tonight.* The lady then disappeared around a corner.

The big orange sun decided to nestle on top of distant oak woodland and I snapped away dozens of photos. Every now and then, a thin strip of pink atmospheric particles would make itself visible against the brightness of the orange, adding to its already surreal beauty.

Suddenly a commotion happened and below me, an area of field and scrubland was alive with the cacklings of magpies. About half a dozen of them, along with many rooks and carrion crows, piped up, making a racket at the edge of the woodland where a field began. I looked but couldn't see anything for a few minutes. I continued to photograph the sun as it closed in on the more distant horizon, losing some of its brightness and now ranging from bright orange to a warm, mellow pink. My camera was trained on the west. I then spied something. A movement caught my eye just at the edge of the woodland under the barbed wire fence. An animal ducked under the fence and entered the edge of the grassy downland. I expected to see a fox, but it wasn't. It was big and beige. It was walking slowly towards the main track, and as it did, I could see that it was a large cat. It was long in the body with a very long looped up thick tail the same colour as its body.

The crows went ballistic as they took no notice of them but walked onto the track and sauntered along for a few meters before turning off the track back into the long grass. It then turned around and walked back onto the track for a few paces before going again to the long grass. Each time it came out, it seemed to flash a huge sigh stating, "look, I am a puma."

It was so obvious, but the colour was very pale, the colour of dead grass. When the cat turned into the long grass, it because invisible as it was exactly the same colour, but when it came out, it was so obvious some people may have said it was white!

Just then, the lady came around the corner and the cat vanished back into the long grass. The birds had stopped cackling, possibly because of the sudden appearance of the human. She walked over to one of the small ancient burial mounds and sat on it to watch the sun sink below the horizon. Just as the sun sunk, the cat appeared again out onto the track directly behind the lady who was totally oblivious to it and was looking in the opposite direction at the sun. The cat could see this and so summed up the situation. I couldn't believe my eyes and almost felt like shouting at the lady, "look there's a puma right behind you!" But of course, I didn't, and what else I didn't do was move my camera quickly enough. I had to grapple with it from the brambles it was and line it up on the slope amongst the bushes, which wasn't easy, then I couldn't see through the little view properly and I could not see the view on the screen as it didn't do that, so I had to judge where the cat was in relation to my pointing zoom lens and snap away hoping to get something worthwhile.

I snapped five photos, one of which I captured the puma only just, but the photo was bad and the beige shape could have been grass for anyone else knew, but I knew that it wasn't when I viewed them later. The puma was back in the long grass but now not walking as it did before, but in full stalking mode, slowly creeping along low to the ground. There was a roe looking straight at her in the field edge, but another one I hadn't seen suddenly appeared in view under a thorn tree. The puma was in full alignment with this roe doe, and she hadn't seen it. The cat disappeared from view in the grass and the lady was still watching the sun. The globe fell under the land's end and the breeze stopped as it usually did. All went much more silent except for the constant hum and buzz of millions of flying insects. The lady got up and walked back the way she had come. I remained at the spot watching and waiting but saw no more and then about twenty minutes later, I heard a deer screaming once and then all was quiet.

A week later, I searched for any remains of a kill but found nothing but excellent paw prints and two scrapes.

After this incident, I found female puma scats and footprints on a regular basis.

The remains of a roe doe near Wareham, Purbeck, March 2015, in the same field where I found the half-eaten fox carcass. The skin has been turned inside out at the front, including a front leg pulled inside out. As stated before, this feat involves brute force, a strength that humans and most dogs do not have, but a big cat does.

The Road Sitting Cat

In April 2019, I had another quick sighting. I was driving to the main hill near the three counties border when I decided to search an area that I hadn't been to for a couple of years due to my illness, making me unable to climb hills or even walk very far.

I picked a near site and was feeling rather well after a long spell of chronic fatigue, so as I drove down the small lane towards the hill, I turned a bend in the lane and there it was. A large cat just sitting in the road. It was a good job that I was driving slowly with care, as I always do, as I do not want to kill any wild or domestic animals. Any faster than I may have hit it, but it was quicker than me or my car and as soon as I spotted it leapt into the low thin hedgerow.

I stopped and searched for it with camera in hand, but it was nowhere to be seen. It must have been crouched nearby flat to the ground as it could not have had time to run the two hundred yards or do to the woodland edge higher up the slope and it wasn't in any other part of the large low cropped field. Well, that was that, a blink of an eye and I would have missed it, I only judged the sighting to have been under one second. All I saw was a black cat the size of a Labrador dog with strong hindquarters. It seemed to have its tail wrapped around its body and I didn't see it full out as it leapt.

I then thought that I might be able to see it from on high; I climbed a small hill, and nearly collapsed twice. Still I kept going as I had already started. I scanned the lower fields but saw nothing. There were crows making a din much further away. Maybe that is where it went? I searched the many large conifer and oak trees for evidence and found many old tree claw marks from cats climbing them. So the leopards still come here, just as I thought, although there had been more sightings of puma-like animals lately and in the past. The claw marks could just as easily have been those of pumas if they regularly climbed trees to escape dogs or the fox hunts, as many a hunt as flushed out large cats, only for them to run away like lightning or to be treed by the hounds and many people have seen and heard the cats during the chase. I think it is most likely both car species inhabit this region just like most other places and in many places, all three large cat species co-exist alongside each other and all may occasionally climb trees but especially the leopard as it is more adapted to do so.

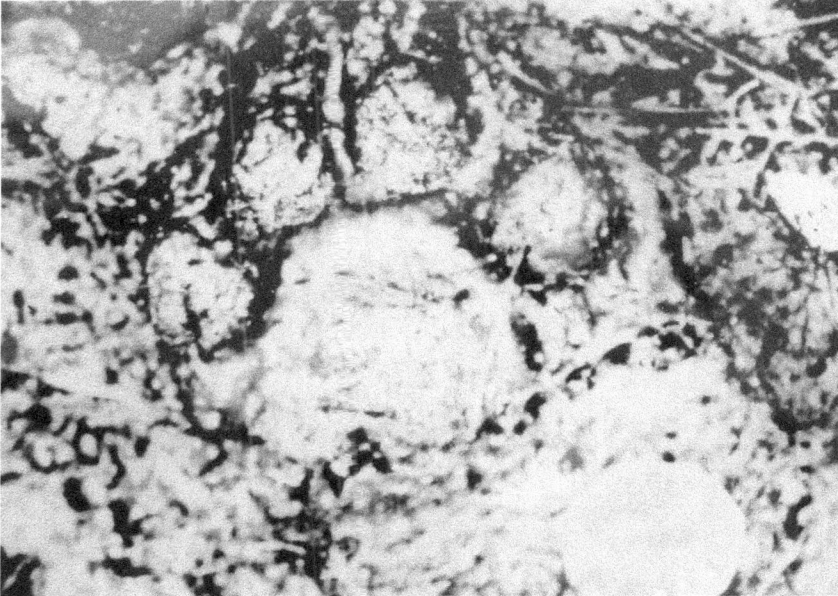

This is possibly a small leopard or Puma spoor from the East Dorset ridgeway. The animal leapt two meters to a dry bank from the deep mud. Lynx has been seen in his area and it is an area of thickly forested hills and many hares and roe deer. There have been more leopard sightings, including a mother with cubs.

The areas of chalk downland in mid-Dorset are renowned for their large cat sightings and I myself saw more than my fair share of pumas, especially in these areas. There are also leopard-like animals and in the same place where two deer stalkers saw a black cat curled up on the bonnet of their vehicle was this typical large cat footprint. What is more diagnostic with this print is that the hindfoot is superimposed on the forefoot, showing a direct register; this means that the animal places its hind feet within or just behind the area where the forefeet had been. Many animals do it; badgers are a prime example, but foxes and domestic dogs and wolves do not. When a cat walks at a leisurely pace, a direct register will usually show. Compare this photo to the two pictures of domestic dog prints. The toe pads are also more pointed and there are no visible claw marks. Sometimes cats pug marks will have claw marks, but they are different in shape and higher off the ground because of the arch in the bottom of the claw.

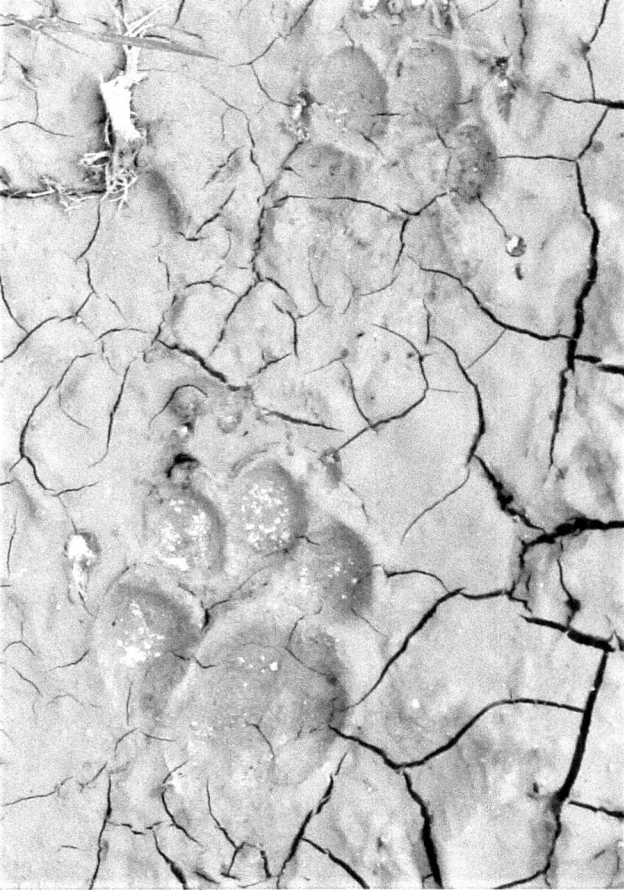

Spoor of both front and hindfoot of a smaller large cat found near the three counties' borders.

There are many times when I have been out and had been aware of the presence of one of the large cats but failed to actually see it or them. I have often had that sense of being watched many times. I have suddenly turned my head to gaze at the distant stand of trees or the hilltop in the natural instinct that I was being watched, and in this game, I have had the impression that there are more of the cats watching me rather than me watching them.

I have often heard and seen the flocks of crows and magpies mobbing something out of my sight range, perhaps by trees, low bushes or a dip in the land; places concealing those whom do not want to be seen and purposefully hide as they had already seen me way before I saw them if I actually did, and to be honest most of the time I know that the cats are watching or have watched and I have been oblivious or often just sensed then scrutinizing me and my possible movements.

The leopard sums it up; he or she sees a human from a long way away and must conclude which way they are likely to go. It takes action either to flee and hide, to sit low and keep still, or in terror, run out in panic, giving itself away!

Fortunately for the cats, the latter idea would be the worst-case scenario and that goes against the cat's behaviour and adaptable traits. I have followed the panicky alarm calls of blackbirds and blue tits, robins and pheasants and partridges in a bid to see the culprit. If it is a fox, then usually I would see it. Birds of prey are easier to see when birds mob a tree, for example, as the owl or buzzard will not put up for it for long and wing itself away with obvious calamity. Stoats and weasels are fast and furious and are hard to see in vegetation, yet the sharp eyes of birds can see them. Birds are so on the case when it comes to predators and especially cats that the cats can barely walk around without attracting their attention of them, especially the corvids. Even at night, these birds seem to be only half awake a lot of the time and suddenly come alive when a cat enters the arena.

I have seen twitching bracken and heather where I was convinced a large cat was moving through it but never actually saw the animal. On three occasions, I saw black fur for a fraction of a second along with bird alarm calls and knew it must have been a large cat. Lighter brown shades could have been a puma or a fox, but when jet black is seen to be manoeuvring around vegetation, then it is most likely to be a black cat as black foxes are very rare and I have only seen them once, but many other weird colour morphs. I have often followed leopard spoor to have it go crashing through the heather and birch trees like lightning but not see the animal, or maybe a glimpse of dark colour here and there if I am lucky, but enough to know that it was one of the large cats that I had been tracking.

These cannot be classed as sightings surely but are of just as much significance or even more! I have once or twice heard the claws of a leopard sliding down the rough papery bark of pine trees as they descend in an effort to escape and hide. Most leopards will always come down from a tree when they see people and then skulk off to hide in a ditch or thick brambles or gorse. They know that a tree is a trap! What's more the puma has evolved to do exactly this when chased by other carnivores such as wolves, bears, etc, this, of course, in its natural state, serves to protect, but now humanity is on the scene. This survival strategy backfires as humans then shoot them! But still, the leopard will always come down from the tree when humans approach, even from far away. The leopard has more white than the puma!

Cause And Effect Without Visuals

Many a time, I have been at a site where a cat was but disturbed it. I may not have actually seen the beast but the after-effects had been obvious. I was with Rick Minter once in a favourite quarry of mine on the Cotswold escarpment near Stroud. I had found the tree wedged roe buck here plainly eaten out by a large cat. The area had several points that stank of male leopard spray.

We decided to enter the quarry from the top rather than the usual bottom road and in doing so, we were quiet. We found an old scat on the slopes on a deer trail, white and grey and twisted like a spring. So we walked further down the grassy slope with small rowan and oak trees, and an animal got up in front of us, only three meters away and I could hear it rushing down the slope and into the dryer leaf litter of the beech leaves as it hit the woodland. As it went, great tits and blackbirds gave alarm calls and we could trace its movements as it went, similar to the Cotswold puma in the driveway.

We came across the area where the cat was sat sunbathing, the area of flattened grass. These are ideal places to search for licked off fur on the ground. I don't think that I found any on that occasion. So although we didn't actually see the beast, we could smell it, and experience the behaviour of other animals affected by it and the marks on the ground where it had lain. There have been many similar circumstances where I have known that a cat was very close and found the evidence but no actual sighting. Of course, most of the time, the ground evidence may not be there, or people will not look for it but still have a close encounter without knowing it! A classic example is the thousands of reports of dog walkers having a growling dog looking at seemingly nothing in the bush when in actual fact, a cat is sitting in there just a few feet from a person. Of course, this sums up the very nature of large cats, especially the leopard, the master of disguise. I have many times seen deer alarming and heard giving alarm calls knowing that a cat was present but not seeing it.

In 2018, I teamed up with musician and ex-policeman Pete Leg. Pete was interested in natural history and loved cats, especially big cats and loved photography. He was interested in local sightings and we both would venture out health permitting to local areas where people had reported sightings. Pete and his wife suffered pet cat losses and many cats in the village of Rockborne disappeared without a trace.

He, and later I, concluded that a large cat, possibly a puma was responsible, yet only a few miles away, a leopard was busy. It didn't take long to find all sorts of the typical field evidence of a small leopard living around built-up areas but adjoining fields and woods.

A man had seen a leopard descending headfirst down a pine tree at the end of his garden and another person had seen a huge black cat in a lane nearby. We found many paw prints of this smallish leopard along the lane and we decided to walk alongside big oak trees. I suddenly felt as if I were being watched and looked straight at a huge old oak and said, "that looks like a perfect leopard tree," and as I pointed to the tree, Pete saw a black leopard quickly scale down the tree on the reverse side to us. I saw a quick flash of movement, I at first thought it was a jackdaw as two and a pigeon was sitting further up the tree; however, it was apparent that an animal with lightning-fast reactions got down and kept the tree in line with us as it departed.

Great tits and blackbirds gave alarm as it ran through the field and bramble thicket into the next field and then up into horse stables area with derelict buildings. We couldn't see it go, we didn't need to, the birds showed us the way. Pete took several photos as soon as he saw it but it went too fast. He was chuffed that he saw it, I'm glad he did. I wasn't bothered by the fact that I didn't see much. We tracked into the field and found a scrape and fur on brambles along the path it took. We found a lot of tree climbing claw marks on most of the oaks around the site and a few hairs on the bark. We later found footprints under a drain under a rack and other evidence of many months. It was interesting to note that the leopard only descended after I looked and pointed to the tree otherwise, it may have been there all day and for the rest of the day, not bothered by the few people walking along the lane. Eye contact is a giveaway to an animal and that usually means time to fly!

4

TO CATCH A CAT ON CAMERA

The Attempt At The Impossible?

With apparently so many large cats wandering around, it is obvious that many folks would set about trying to capture them on camera. We are talking about the most elusive of all animals on Earth. People often quote, "how come there is no solid proof, such as photo or video evidence?" That is a valid point and is one of the enigmas of the subject! They seem to evade all attempts to capture them. How then can the wildlife photographer manage to capture one of these beasts? One would wonder why it has not been done already.

There are thousands of naturalists in the fields, woods, moors, mountains and hills studying all forms of wildlife. It is more than bizarre to think that no one has even accidentally come across a cat and managed to snap a photo of it. Well, it has actually happened, but the photos are not too good. They are grainy and the subject is far away. They are not like the typical photos we find in wildlife books or magazines or the amazing footage captured on television documentaries. There have been many attempts to capture the animals on mobile phones and video cameras by folk who have just happened to come across them. This is the main form of footage and still photos that we have of these animals. As far as I am aware, there has not been any general naturalist hell-bent on getting proper photos or film of these animals to date, except a few open-minded people that are into the subject, some of which are serious naturalists.

It takes skill and dedication on an immense scale to study these cats enough to take photographs. The people who have been at the right place at the right time have not been naturalists but people with other interests; otherwise, there would be many good photographs. The nature of the animal's behaviour makes it incredibly difficult and unless there is a female with dependant cubs, with a known area, it could well be near impossible without many months of serious research.

The most obvious way of capturing images of such elusive animals would be by means of trigger cameras positioned along animal or game tracks or where animals are known to walk or hunt. This is a now popular attempt at trying to prove that these cats exist. There are many cameras as I write now up at locations all across the UK and Ireland, also in North America and Australasia.

In Britain, there are hundreds out capturing images of all manner of wildlife. There have been only a few obvious photos of large cats. There are many possible ones, often taken from behind rather than head-on. Many of the alleged footage or stills are of domestic cats or other wild animals. One or two are plainly big cats and most possibly in the form of common leopards. There are a few obvious pumas. The cameras that have captured these cats have been put up by people who are very interested in the big cat subject and usually have some knowledge of natural history. They are mostly members of some of the organisations devoted to the subject.

There are also lone workers who do not want to be part of any large cat organization, who may liaise with like-minded people or do it totally alone, having had bad experiences with the sometimes bad attitudes of some researchers. Trigger cameras have picked up large cats accidentally when being used to research other mammals or birds. Many readers may be aware of the sensational news that forestry commission employees picked up large cats when doing deer research. It would seem that much footage has been purely accidental and coincidental and the operators of such systems have been pleasantly surprised or shocked.

I am sure that much footage of large cats captured in this way has not even been looked at, or even if they had, the operators thought nothing of it or had no interest in the subject material. There are security cameras everywhere, in car parks, industrial sites, town high streets, country estates, farm yards etc. If there were a National campaign for all unusual animals caught on camera to be handed over to any authority, then I am sure that we would be innundated with amazing footage.

I acquired two trigger cameras to use by a friend and enthusiast, Dave Mitchell, who was interested in the subject. They were both wild view cameras, old style with built-in flash. I first used them in an area of chalk downland on the mid-Dorset ridge, amongst forested hills and valleys. The area was only three miles from the spot where I had encountered the quarry pumas. I had found the spoor of leopard in the area, and there had been several sightings of big black cats, including the classic car bonnet bin bag account. Two official deer stalkers came back to their vehicle without even sighting any deer strangely to them. On return to the off-road vehicle, one of them noticed what he thought was a bag of rubbish dumped on the bonnet. As they neared, the bin bag, un-wound itself and skulked off into the forest! It was an area rich with roe, a few fallow and sika.

I baited one camera with the carcass of a young sika stag and was amazed to find that no animal touched it. No buzzard, raven, fox, badger or cat. Only the flies celebrated as a hoard of maggots consumed it to the earth. Why had no other animal touched it? Surely foxes would have been hungry enough to take a few morsels? I later realised that not many animals take carrion in the summertime and in areas with much food during the summer months, not even foxes will risk taking meat from a suspicious dead animal. No cats in this area were captured on the trigger camera.

After six months I moved them. I used them both for over a year in West Dorset until one was taken possibly by poachers. Poaching is a big and serious business where I live and the people who do it are usually nasty uncaring people who are out to make money with total disregard for animal welfare. They are reckless in their approach and use any methods to wholesale slaughter the deer. There is so much poaching in the countryside around Dorset that I feel so saddened by it. Deer feet and heads can be found behind any convenient field gate in the countryside. These people usually use modern gadgets in their quest to go unnoticed and use night vision goggles and scopes as well as silencers on rifles (some poachers take animals just for their own consumption and of course, I would not be too quick to criticise that). Many folks turn a blind eye because it is claimed that deer numbers need to fall. Local police and other authorities do very little to prevent it or take offenders to court. The crime on this scale affects all wildlife, not just deer, of course.

In an otherwise quiet area of forest or heathland, people roaming at night is not good for animal welfare. Large cats are also affected. The problem with the older models of trigger cameras was that the flash could be seen from a long way off, thus attracting curious minds! Whoever the people are, indulging in this activity regularly see big cats, especially if they are using night vision equipment. Some reports have come from poachers or people legitimately using modern technology. It was known that poachers were using the fields where I had my camera up on a nature reserve, one of Britain's best in regards to biodiversity and natural woodland cover.

I actually put it on an oak tree near to the ground with an animal track in view leading under a barb-wired fence into a fallow field. I chose the place because I had found several mammal whiskers wrapped around the fence. This is a story in its own right and I will refer to it in a later chapter on hair.

I left that one for two weeks before excitedly changing the data cards, and swapping them over with clean ones. Would I have a photograph of a leopard or a puma? The first moments of scrutinizing all the images were exciting, to say the least. Badgers, foxes, roe deer, squirrels, cattle and wood mice were the only beasts captured. I acquired no photographs of any cat-like mammal. Perhaps I had to wait for months, perhaps the cat only passed by every few weeks or so? I only managed to check it twice before it was removed, with no trace of it anywhere.

I placed the second camera about half a mile away among some woodland game trails. I did not leave the cameras out there for a long time period as the early cameras used up a lot of power and I wasn't too sure how long a time would be adequate. It was too far to travel every two weeks to check them, so I ended up leaving them for a month at a time, but on doing so, I would find that the batteries had died. I had had several reports of big cats in the area and with a large herd of fallow deer and many roe and wild boar, I naturally assumed it to be a hot spot for cats. It certainly was as Indeed I had found spoor in path mud and the remains of deer consumed by large cats in the typical manner. I did have many blank shots; that is to say that an animal had triggered the camera but it had moved too fast to be captured.

After eight months of nothing, I decided to remove the remaining camera and put it nearer to home. I put it on another nature reserve managed by the Herpetological Conservation Trust, now renamed The Amphibian and Reptile conservation. This is an area of heathland with natural native and pine woodland. The area had the largest amount of no official reported large cat sightings within the county. I wasn't surprised, to say the least, because the area has the largest roaming herds of wild deer in Britain and the biodiversity is the richest in the whole of Britain. I had seen large cats in this area and it was one of my main study areas. I felt convinced that I would capture the animal very soon. Little did I know that I was to be led on a wild goose chase of frustration?

I decided to take the camera away for a while as kids were using the area as an off-road bicycle track. The cameras went back in the wintertime, when there was less activity from the kids. I, at first, did not bait it and got only photos of common wildlife, lots of deer, fox, and badgers. I put a roe carcass in front of a camera here and a fox ate most of it. Another carcass was placed in the same spot and a buzzard ate the whole lot. Hundreds of shots of one buzzard in all positions eating the whole thing, and at the last moment, a large fox comes along and lifts the remaining skeleton and takes it away.

Another roe carcass was placed there. It was springtime and I was rather optimistic as a roe skin had been found under a bush just two hundred yards away. Only one leg remained on this young deer. I knew it was a large cat that was to blame.

The following week an ecologist worker told me that she had seen the remains of a carcass up a tree. I went to the spot and found a leg underneath. It may have fallen, but it had been clearly licked clean. I placed a deer body nearby with the camera on it. It seemed to rot away without even a fox attempting to take any. I was surprised foxes did not attempt to eat; why not. I left it for two weeks and returned to find it gone. "Wow, I must have it." I chuckled. I searched around but found nothing.

When I got the camera home, I was shocked to find the batteries had died, perhaps just days before the bait was taken. I was gutted. How typical. It was sod's law. This law was to repeat itself time and time again. It was as if I was not to get a photo of the cat. I put the camera back and left it again for a few months without any bait and only got the usual wildlife, but one shot shows a large furry head right in the frame. It was overexposed, so just looked like a huge head. It could have been a cat, but the ears were not in view, but there may have been whiskers. Anyhow it was not obvious as to what animal it was, so I binned it.

During the summer, the camera had been baited three times with deer. Once I put the body of a young sika stag by it and left it for two weeks. I returned to find it gone and again, I thought that this was it. I searched high and low until I found this one. Half-eaten, it was hidden in some reeds on the edge of the bog nearly one hundred yards away. It could only have been dragged by a very strong animal. Again, when I took the camera home, I realized that it had become faulty and had stopped working. How strange. The bait was removed after the camera had stopped working. I realised much later what was happening. Two new modern cameras (little acorn) were passed on to me by an independent documentary maker in a bid to capture a photo of the animal. I placed them in various places, sometimes in the same place as before, other times in what I thought would be better sites and usually baited them with a deer carcass. On three occasions (at the time of writing), the bait was taken when either the batteries had failed or when the camera did not operate for short periods of time. For example, we put a carcass by the camera about two meters in front of it. For days just buzzards, crows, magpies, a few badgers and mostly foxes seemed to pick at the carcass. When it came to the foxes, they were always very careful and at first, they would not go to the carcass but linger around the area. They were clearly very suspicious and some did not attempt to eat at all. Others seemed to pluck up courage after several visits. On no occasion did I ever see a badger eating from a deer carcass or any other bait in the form of meat left by a camera. The badgers used to walk by as it was often on their path. Usually, the presence of a carcass would put off a badger from walking the path. This is evidence to suggest that most badgers are not carnivorous or take carrion. But what was plain to see was the fact that certain mammals actually see the infrared lights on the display on the front of the camera. Time and time again, a fox or badger would amble towards the camera and as soon as it started recording, the animals saw the lights and turned on a sixpence and fled. Some never came back; they were too terrified.

This opens up debate on what actually a mammal can see in regard to coloured light. It also should open debate as to what animals sense as it may also be linked to electromagnetic fields and if so, then no wonder that large cats keep clear of them. This is hard to believe, but those long thick whiskers are there for a reason and it may be that these cats feel them meters away, or at least see the lights on them or hear the motor running or a combination of both. Cats can hear frequencies that human ears cannot! On both high and low scales. Why else would the cats strike only when the cameras were not working? This happened three times with the new cameras. Clearly, the cameras did not capture everything that went on. The cameras did not run continually whilst there was movement in front of them.

This was apparent later when other people experimented by placing cameras alongside each other. One would record an animal that the other did not, or only some of the activity and would usually only kick in when the animal was in mid-frame or even beyond, whilst the animal was walking away from the camera.

We thought about putting them overhead and not face on so that the animals were less likely to look up and see the light, but in the very darkness of the wilderness areas in which I was putting them, even the small glow lit up the whole area. Finding a location where there was a bit of light, such as a street lamp that was safe and where I had permission and where a large cat would pass, was near on impossible.

There were cats that roamed through villages and small towns in the early hours, and a few cats went into the large towns by way of the railway line. Cats often use railway lines, used and disused, in fact, the used ones are preferable as they have no humans or dogs using them and the cats see the lines as security. Possibly, the whole rail network of the UK has been walked upon by large cats. They are used as highways from A to B on a regular basis, some cats use them habitually and some use them as a safe base to lie up during the day.

Some of the city cats may do this, and many a country cat certainly does this. Some cats will find themselves on a railway line when looking for new territory, such as youngsters who just left their mother. They may use it only for a while but will remember it. They get used to trains and older animals will know about the electric rails and class the line as its home.

Railway lines are also home to other animals. Deer and foxes use them as highways also and so do badgers which often have their setts on the banks. Most railway lines have ample cover and are usually wooded at the margins. This cover is essential for shy animals, serving as corridors linking areas together, in turn allowing wild animals to travel undisturbed from one end of the country to the other. Deer stag carcass was placed by a camera, only to be eaten by buzzards and foxes many times, and many times a carcass would just disappear without a trace. Occasionally the carcass would just rot with nothing touching it at all. Since I wrote this section several years ago, we have made advances in technology regarding game cameras and now there are superior models that have a kick in speed much faster and one can notice the difference.

A very big cat of some description had walked past the camera before it triggered the infrared device. It was only caught because it was walking slowly. Any quicker, then the camera would have triggered at its exit and nothing would have been seen on film. Many a cat swiftly walks past, creating just a blur, or setting them off but not capturing the animal. The gorse bush between the cat's rear and its tail is fifty cm in height. The tail is head on, hiding its true length. The cat is deep in body and looks like a puma but could be something else. This, I believe, was a puma.

Sharon Ramsdons's game camera photo of a large black cat scratching on a small tree. A catnip bag hangs to attract it. Note the thick short forelimbs and ears positioned far apart and held far back on the head! This could be one of the typical stocky hybrid cats, or it could be a young leopard or puma.

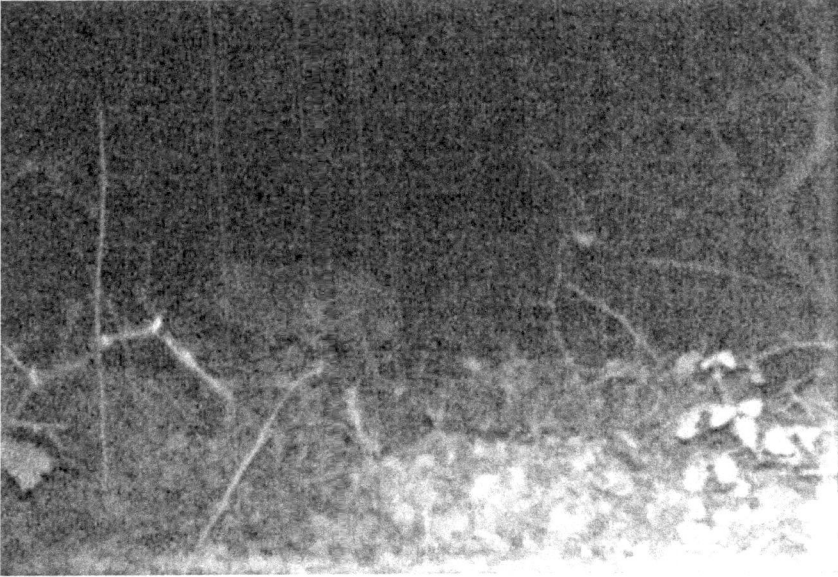

Another of Sharon's trigger cameras has captured what looks more like a typical large cat in shape and size. The cat is walking from left to right and has its head tucked down in a typical large cat posture, its tail is either very short or long and curled up at the end. There are spotted areas which relate to a tail but may not be. The animal could be a lynx. The spotting on the animal is consistent with the typical spotting of a leopard, the pixelation of the general photo does not help, but one can see the differences between real spotting and pixelisation, several attempts to erase the dark botching have caused the animal to be viewed in a different way, and thus resulting in some people claiming it to be a fox or even deer. To me, there is no resemblance to either, as some things are not right such as the leg alignment and the shortness of the neck and body. Even a fat dog fox will look longer usually. It also seems to be smaller than at first anticipated and although there is much interest in the photo, people are divided as to what it really is. There have been many reports of spotted leopards in the Eastern end of southern England, especially Kent and Sussex, with odd reports from other areas of Britain. Spotted cats would be much more difficult to see as the black ones stand out far better, especially from a distance, and so would be reported fare more frequently than normal spotted leopards.

The small female Poole town leopard. This photograph was taken by Mrs Beatrice Munday of Poole Right in the centre of the built-up areas, she saw and took a few photos of the animal as it hunted squirrels and was mobbed by crows and magpies in a tree behind her house on the grounds of a primary school. No children or staff members were present as the schools were closed down due to coronavirus lockdown and so the leopard was at ease, although it was just gone midday. This leopard was regularly seen around the same time of day in the vicinity, including on the grounds of Poole hospital and parks in a small area. There is a magpie above and to the right of the cat, the bird seems to be in the foreground a few feet in front.

If areas are flooded with cameras, then the cats may be put off from entering the area, as we have sussed out that in some areas where cats would often leave physical evidence, there is no more after a camera or two is put up, and if many cameras are put up, then no more evidence is found in that particular area suggesting that at least some of these cats may be very sensitive to them and actively avoid them.

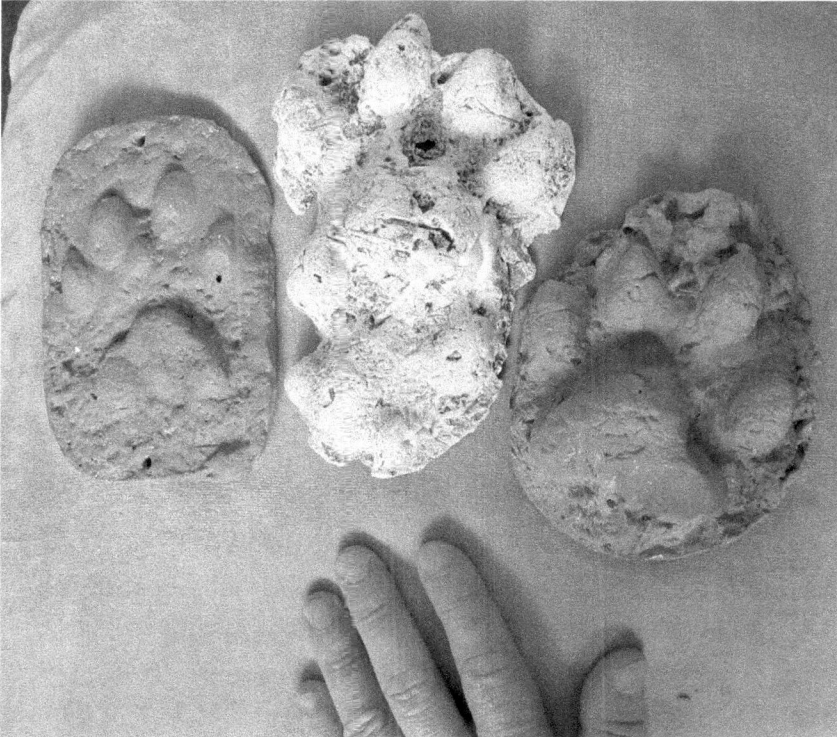

I took plaster casts of some of the mud spoor. One cast shows the direct register that cats display when walking at a leisurely pace; that is to say, the hindfoot is placed just below the place where the forefoot was placed hence a double footprint. The hindfoot is quite different from the broader forefoot and the toes are more angled.

Trapping Cats

Catching the cats on camera seems hard enough, but what about catching a cat in a trap, such as a wire cage or a leg hold trap? All species of cats are capable of being trapped in these ways, just like many other mammal species, but cats, being very intelligent and wary animals, may avoid it, as part of the cats' success at living is to be very careful, but in another way, the cat may be more likely to be caught because it is curious, and we know what curiosity did to the cat didn't we?

Cats have been caught by accident in traps targeting other species. Large cats seem to be less likely than their smaller counterparts and in fact, domestic cats are more likely to be trapped than any other cat type or species. Domestic cats are very curious and poke around in places, often getting themselves stuck in confined spaces such as wall cavities and boxes, drains or up in trees.

Hunting cats are even more likely to check out odd places and a hungry cat more so. Domestic cats are always attracted to carrion or anything smelly! Baiting a hungry cat is the only way to capture one, but even then, it can be touch and go and many other animal species will be trapped before your leopard enters. Leopards and puma are regularly trapped, especially by leg-hold snares, as these are far more likely to capture suspicious cats than cages.

In many countries, cats are snared but usually by poachers, so they are not harmless but cut into the cat's ankles or neck with a wire trace, often cutting into the flesh. Snares are nasty and cruel and do not discriminate. These days, humane leg-hold snares are more often used not just by hunters but biologists when studying them. Many other animals and birds can be caught in snares meant for another target species. People have tried snaring large cats in the past and no doubt some at least have been caught in fox snares or even rabbit snares. Some larger cats may be able to pull themselves loose, others stay put and suffer the consequences. Even humane snares cause trauma, injury and often death.

There was a well-known report of an illegal gin trap being found along with the remains of a large cat-like a leopard in England somewhere. Of course, snaring of any cat species is illegal in Britain and I would never encourage anyone to do so. In fact, any snares of any kind are inhumane. There is no such thing as a humane trap, as all animals suffer when caught. All traps used in Britain need to be checked at least three times a day, regardless of what the quarry is. The idea of trying to trap some of these large cats is now old as it has been tried here in the UK before. Under the right circumstances, though, it could happen with minimal damage to the cat. The best way is to bait a tarp with a small secured piece of meat that is taking able whilst the main bait remains out of reach. In eating the small amount, the cat can take in a drug positioned within it. If people are waiting nearby with a live view camera, they could then home in and take the sleeping cat, take the data, put on a collar and put it back without the cat even knowing what had happened.

There are reports of gamekeepers using large traps possibly to capture large cats to dispose of. I came across a huge trap set by a keeper which could have held a puma; it was at a site where puma cubs had been seen on the edge of Bournemouth and Christchurch. I wondered!

The late Quentin Rose lived in Tewkesbury, Gloucestershire. He studied zoology at Cardiff University and later lived for short periods of time with Canadian Indians, as he became interested in trapping animals. He thought that it could become a full-time business for him, very unusual, full of adventure and excitement; it sounds like a good idea in principle but in practice, not so.

After over a year of bureaucracy regarding the alteration of a well known American spring trap called the Aldridge trap and getting the licences and go-ahead from the UK government bodies, Quentin did not catch any large cats. He did catch huge domestic cats, though, and many other none target species. By the time Quentin got all the go ahead's to do what he wanted to do, the cats had moved on elsewhere. Quentin was hitting on a new subject for the British government and a tricky one at that. Some officials were not happy with what he was doing, and others were. Most people were actually suspicious and did not think it was a good idea.

Quentin's idea was to capture the leopards and pumas and re-home them back to their countries of origin after being kept and scrutinized by British vets and zoologists. Quentin, at first, may have thought that there were just a few escaped pumas and leopards, but as he did more research, he started to realize that there were many of these alleged escaped cats and soon realized that they were breeding.

Did Quentin think that he was going to capture all of these animals and return them? Well, if he did, he had another thing coming because he later must have realised how impossible that would have been, not just because he didn't manage to trap a single large cat, but to do it properly would have meant knowing where all the animals were and to judge their next move. Quentin would receive one sighting of a large black cat from a village somewhere, for example, one morning, then during the afternoon, he would get another report of a brown puma-like animal just a few miles away.

Quentin did not like the idea of leopards living in Britain as he was well aware of their potential as man-eaters when wounded. Quentin was once attacked by a captive leopard and so maybe held a slight grudge against them; certainly, he liked to over sensationalise their potential.

Yet pumas kill people in the Americas, not quite so many as leopards in other countries, but basically, they are both capable of killing and eating humans if need be, but of course, these are always exceptions to the rule. The leopard does slightly have the upper hand when it comes to strength and endurance and wit. Quentin was always against people shooting the large cats and rightfully so, but surely one should not discourage it on just man-eating grounds, but on all grounds.

I do not like the idea of capturing any wild animal and placing it in a cage or any unnatural holding. Animals suffer hugely due to stress when captured, especially wild cats. Leopards are known to break their canine teeth trying to escape from metal cages and snared leopards have bitten off their own feet to escape, something that many animals and birds do out of desperation. When a large cat has its teeth broken, it often cannot kill efficiently and may starve to death due to the consequence. If not, then they may get ill due to infection and then die. The ones that survive may start to hate man even if they did not before and they can take revenge and become a man killer. This has happened before. The only reason I would ever recommend it as if in the advent of conservation bodies wanting to do research. If a radio collar was to be fitted and blood and fur samples were taken, for example and the animal to be set free, this would justify any risks, but even then, there should be a radio trigger to announce the cat's arrival so that a tranquillizer can be administered as soon as possible.

On some reports from cougar trapping in America suggest that some of the animals die of stress and it disrupts their behaviour and mental health. Leopards have been known to attack people at trap sites. They can recognise an individual's face and turn on that particular person who was responsible for doing something such as poking the animal or shouting at it or shooting it etc. A caged leopard is often a very dangerous one. The potential to harm both people and cat are very great. One would need a reason to trap and an idea as to what to do with the animal once it was caught. Some zoos and wildlife parks have often stated that they would take in any such animal. But the fact is, is that most of them are now wild animals that are born from several generations of animals after they escaped or were released from captivity. Why put a wild animal in prison? People with these sorts of ideas must be control freaks. We do not need to do that.

The UK government only ever licensed Quentin Rose to trap exotic cats and some other mammals in Britain, as far as I am aware. I am sure that they have done it themselves, but governments are always corrupt and do not abide by their own laws. Maybe other forms of traps have been used by authorities, as most are aware of the fact that we have naturalised cats in Britain, even if they plead ignorant.

Shoot To Kill Policy

If it is proven that any one species of large cats needs to be eliminated, then it will be a very difficult task and will need money that governments do not have. They need to prioritise surely, and big cats have not been a priority really in the past, but now, as numbers are very high, many bodies of people are being advised to shoot to kill any large cats. This is very irresponsible behaviour and should not ever be encouraged, especially by farmers or other gun-happy idiots. If indeed there needs to be an elimination programme, then it must only be done by a trained special arms unit. They will never be totally eliminated anyhow as there are so many of them and in such wild places as well as cities full of people. Many cats will never be found and hideout. It would take over a hundred years to kill them all, even if they put all their efforts into it. In the end, poison would be used as an alternative as it would be just too dangerous to shoot rifles in some areas. The laws would have to be broken with so much bureaucracy, it would be a nightmare and it would not be done right and a whole cock up would probably happen due to idiots.

Many cats have been shot and killed, but many more have been shot and not retrieved! Food for thought, isn't it? I lost count of the stories that I was told about lampers shooting cats only to see them spin in the air and disappear into a thicket, never to be seen again. It is an offence for anyone without a special licence to even attempt to shoot one of these large cats. It is also illegal to shoot domestic cats and many of them could be killed due to mistaken identity and if domestic cats are very big, I'm sure that they would be shot and questions asked later. The thing is that many people keep large domestic cats as pets. Many shooters may think that one of these extra-large moggie hybrids may be a leopard and shoot it. It is illegal to shoot some person's pet animal unless they are caught in the act of doing damage, such as farmers with livestock. Only then would it be legal to shoot a large cat.

As I have stated elsewhere in this book, large cats are not the same as deer. Cats have stamina and can take a lot of pain. They rarely die easily and will do all they can to escape and die in dignity elsewhere.

I have also added a section on man-eating within the UK because it is very important for people to know about it and the causes. People have been eaten by large cats in Britain and it may be that at least one of these leopards that have been known to do so was injured due to an idiot shooting at it, or it may be because it was hit on the road by a vehicle or due to another kind of illness. The fact is, is that people who shoot at large cats could be committing manslaughter if that cat, due to its injuries, had to eat people, and the shooters were aware of the possible consequences. I personally do not wish all the leopards to be killed just because there is a potential for some of them to kill humans. Dogs kill many people every year, cows kill people every year, bees kill people and so on. We cannot keep killing animals just because there is a small risk of them killing us. We have to die and it is because of all of the problems within our sick society that causes overpopulation of us. People go way over the top and are brainwashed into believing that humans must have no risk, no death. Why? We need risks. We need something to be afraid of. We need some people to die to act as a deterrent also. We need some people to keep away from the only wild places we have left. We need not so many people walking out at night disturbing all manner of wildlife. We naturally need our minds to be shaped in a well balanced natural way, and that involves having something to be slightly afraid of.

This photo shows what could be a lynx or puma. It is thin in the body showing pelvis bones close together and with a deep body not tapering like a canid. The tail could be short or long but curled to the side. It is lighter on the underside. Roe deer have no tails and sika and fallow would be larger and higher off the ground with a large thick backside.

The animal in this game camera photo is smaller than a sika deer with short thick legs and a paw, like a large cat. It could be a puma. Both this and the previous photo were taken from cameras placed next to each other on heathland and forest with lots of large cat activity. Wareham.

At dusk, a leopard at Creech lays low, watching sika hinds. The cat's movements could be traced as it edged its way around the woodlands by crows and deer giving alarm calls. The cattle took no notice, but the deer were on edge. The cat was frightened off by a farmer shooting a starting pistol to scare away deer, but it was seen later chasing deer a few fields away.

A closer view of the Creech leopard from the above photo.

A jungle cat sees me in my car and runs off into the thicket on the edge of Poole harbour.

5

BONES

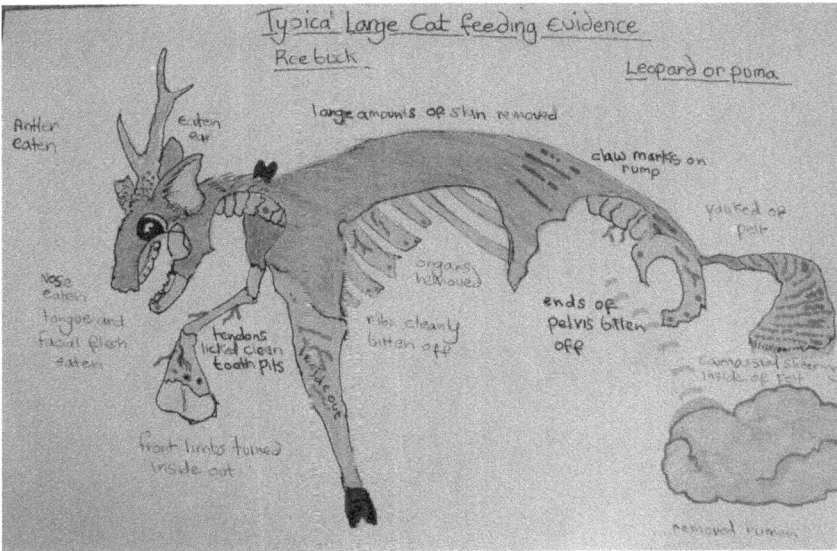

A sketch of typical hallmarks of a large cat feed.

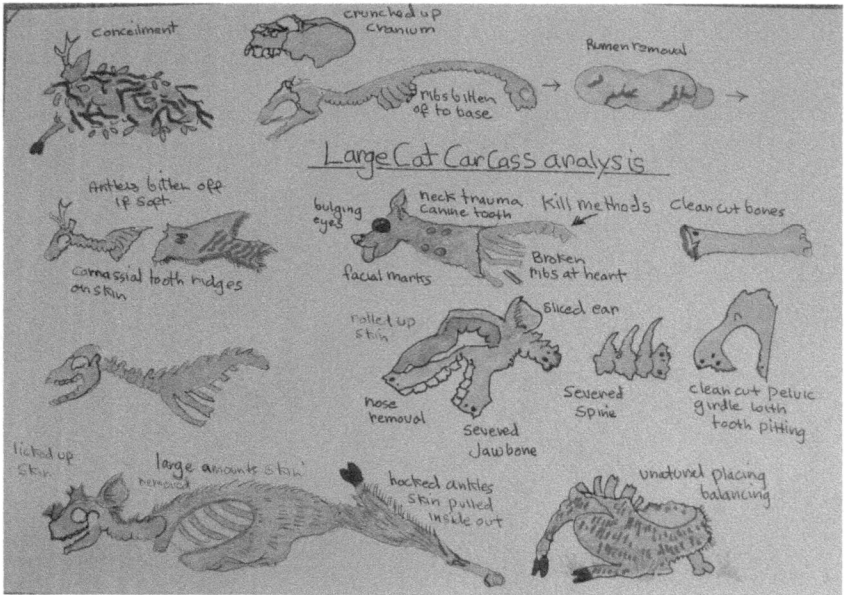

A sketch showing hallmarks of large cat eating.

The leopard of growling corner had the remains of at least sixteen deer carcasses at his HSFP (Habitual safe feeding place).

B ones, bones. Everywhere bones.

Bones in the bush. Bones on the beach, Bones by the lake, Bones under the shed, Bones on the heath. Bones in the deep woods, Bones in the ditch, Bones on the road, and bones in the tree.

There were so many bones all over the place, mostly deer bones. Odd skulls, femurs, scapulars, often on their own, or whole carcass remains. It would take time to suss out what animals were responsible for eating the carcass if indeed they did, and what was responsible for the animal's death. Sometimes it is very easy; other times, it is very difficult.

In an area that has huge amounts of wild deer, there are inevitably going to be deaths. Wild animals die for a number of reasons; one of them is not old age. It is very rare for a prey species to die of old age, but rarely of age-related illness or starvation.

Now that we have the apex predator doing what it should be doing, most animals die before they suffer from age-related problems. I have looked at many bones over the years. I have found so much evidence to suggest that large cats mainly take the less healthy deer. They keep the numbers of deer in check by not just taking the diseased individuals or those that have broken legs or are lame, but they seem to prefer young female deer.

Eighty per cent of the cat killed carcasses that I have found are sika hinds less than two years of age. This means that they are targeting the animals that give birth. Naturally, there are more females than males, as sika like reds are herd animals and a male will have a harem of females which may be over twenty, although the ratio is not twenty to one, but maybe five to one. If stags are taken, they are usually under one year old. The cats would seem to proffer the stags as scavenged meat from roadkill or other sources. I have only found two mature sika stags as cat kills and about twenty odd young stags under the age of two. A few road victims might have been killed by cats if they were unable to themselves. I am not sure whether it is just the fact that there are more female sika deer around or if there is another reason for most kills to be of hinds. Female deer do seem to be more approachable, and the males seem to be naturally on the lookout. Females with calves are even more so for obvious reasons.

Male Bovids have evolved to watch over their females at certain times of the year. They also are prone to be pumped up with testosterone and so are more likely to put up a fight, and with sharp antlers, they can do damage to a predator hell-bent on killing it. Most leopards I have seen on wildlife documentaries seem not to be fussy about whether or not the deer they are hunting is male or female. They are opportunists and take what they can. Perhaps if they are spoiled for choice, and then they can be choosy. Male deer are also much bigger than females. The female sika is manageable in the sense that they are always of a similar size and are easier to drag. A mature sika stag is massive and I have trouble just pulling them a few feet or so on my own. A female sika, on the other hand, is never too big to lift. Also, it would seem that sika hinds are more prone to deformities such as bent nose conditions or have jaw problems. Some specimens that I have picked up also have uneven teeth.

In some places, hoards of bones gather up as a leopard or puma has a favourite eating place. They can carry a carcass for miles if they wish to eat at leisure in safety. Many a time, a cat does not eat its full potential because it is so often disturbed, and so the cat must hunt again. It makes sense for a cat to have a safe place where it knows it will not be disturbed to finish the whole carcass and make the most out of it. I knew of two areas where bones of at least four sika deer and up to eight had accumulated over periods of one or two years. In all areas, the build-up has not re-occurred, meaning that the cat either no longer exists or has found a different territory, possible moved on by another cat of the same species or from human disturbance.

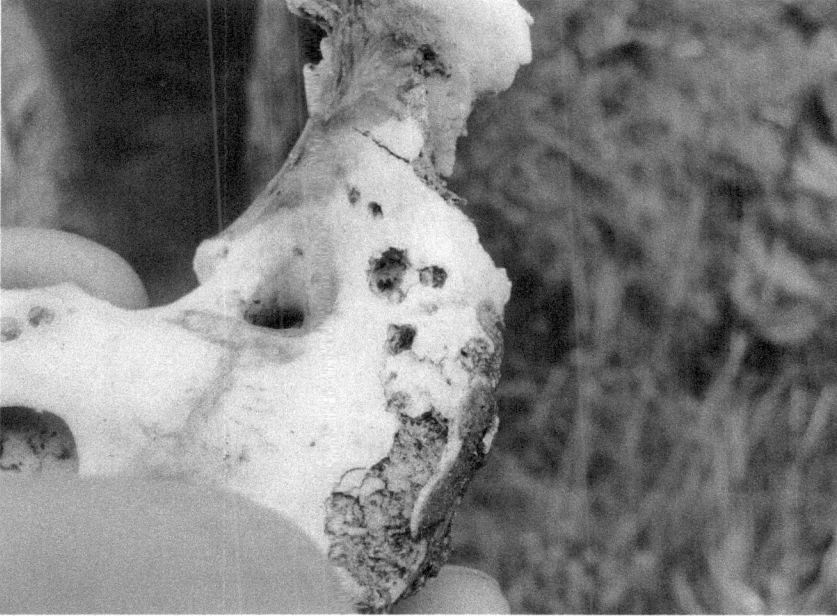

A pelvic girdle from a sika deer was found whilst we were on Dorset heathland at the Dorset cat conference. Good tooth pits are clearly visible within.

Carnassial sheering from a sika pelvis.

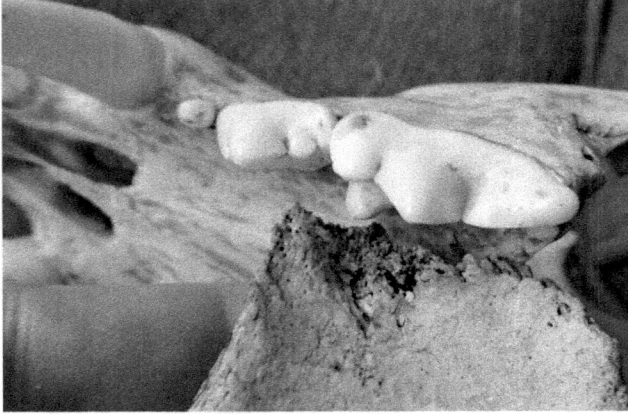

It is a rare occurrence for all cone tips of the hind carnassial teeth to touch bone at the same time as they are at different angles, so it can only leave a mark on the bone surface is uneven, such as a slant or rise, and then it is usually by younger individuals that haven't worn down the cusps. Here is a piece of pelvic girdle that is slanting and the three cusps have then imprinted on the angle. The leopard skull shows where the teeth cusps left their marks.

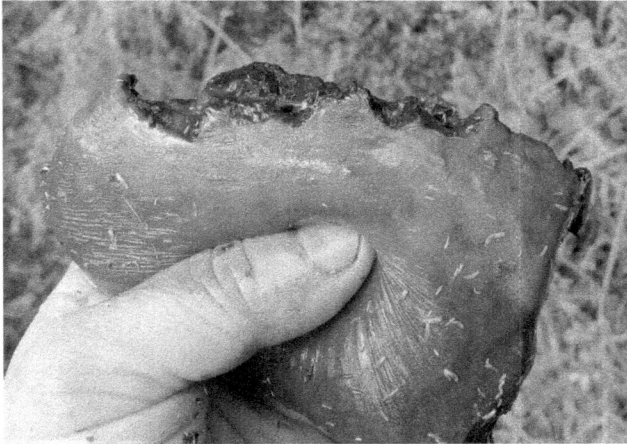

A section of lung taken from a yearling sika stag showing the typical varying angular tooth sheering imprints from a leopard.

This sika hind was hit by a quarry lorry, soon to be scavenged by a leopard and possibly a cub or two.

Large cats often bite the antlers down and leave great imprints, such as this triangle on a sika stag antler. The surface is uneven, allowing for multiple cone trauma.

Other safe feeding sites (SFS) have been found often in thick woodlands, bomb craters, dried out ponds and river beds, lone thickets surrounded by fields, fenced-off compounds, energy places such as electricity places, out-of-bounds areas to people like, industrial sites, building sites fences off quarries, railway lines, school playing fields even. I can mill about in any area of heathland or forest where I know cats to be active and find an average of three deer skulls every time in certain areas, along with other bones. I have often found remains of more than six or seven different individual prey species within one SFS.

Assessing A Carcass

The assumption that a large cat had been responsible for stripping the carcass is detailed. There are some characteristics that are obvious; others are not so clear cut. In my studies, I was surprised many times just how certain species of common wildlife can eat a dead animal in a similar way as to that of a cat and the amount of flesh that a small animal can take. Unless one has a trained camera on the bait at all times, it can be easy to come to wrong conclusions. Just because a deer has no flesh on its bones and seems to be quite clean around it does not mean that a big cat has eaten it.

Some bones are old, many years old, but in many circumstances, one can still come to a conclusion as to what had been eating them. The main and most common scavengers that we have in Britain are red foxes (Vulpes vulpes). Foxes are hunters but are opportunistic and will take carrion, not all the time but if they please and if they are hungry. I was at first stunned to realize that many foxes will not touch a carcass, especially if it has been touched by human hands. Even just dragging a deer carcass by its hindfoot a few hundred yards from a road where it was killed by a vehicle is enough to make most foxes suspicious. As you will read later, foxes are not the blatant scavengers we think they are. Cats, on the other hand, seem to be even more suspicious, and they scavenge much more often than may have been previously thought. Badgers are not so much carrion eaters and will most of the time ignore a deer carcass unless they are particularly hungry, especially during cold winter periods. If their staple diets of worms are not available, then a whole colony of badgers will often reduce a deer carcass to nothing overnight. Most of the time, badgers completely ignore carrion but some individuals actually make an effort to detour around a carcass wanting nothing to do with it as though they are disgusted by it.

There seemed to be a common order as to how a carcass disappeared after I had placed it in front of a trigger camera. I have placed approximately 70 deer carcasses in front of game cameras, and many more carcasses just moved to a likely area or any areas outside my study areas. Many times I have placed part of a carcass in front of cameras also. I often take cuts of meat for my own culinary purposes. Many times the haunch and back flesh had been removed and then the rest of the body with hide still attached or removed. This also shows up the red flesh, attracting birds. I rarely placed small carcass remains or even small deer in front of cameras. Nearly all deer carcasses were of at least half-grown animals but usually fully grown sika hinds, sika stags up to two years old, fallow hinds and bucks up to one year old, or roe adults of both sexes. Only two carcasses were of muntjack. I did not place any carnivore subjects in front of the camera (I should have done).

Firstly, the dead deer would be placed, usually day fresh on the spot. About a quarter of all road-killed deer I picked up and placed in front of cameras were older than a day old and a few were over a week old during wintertimes. All bodies were basically whole and not mashed up. Crushed skeletons and flesh hold no good media for finding typical eating habits of any animal. Bones are broken in a typical way after a vehicle or two has hit the animal, but sometimes it is not easy to tell the difference between the crushing teeth of a large cat and the break of a vehicle impact, but usually, one can tell. Often there are both to contend with.

Here is an eaten carcass of a roe buck at Eggerdon hill, above power stock common, west of Dorchester. This is an area that has had many sightings over the last twenty years. The area is remote with few villages. There are large tracks of natural woodland with fallow, sika and roe deer. To the north and West, the land is generally open farmland with smaller areas of the forest, but the further West one goes into the county, the hillier it gets, and so more sightings are reported, especially from the farming community. This roe has the marks of canine teeth on its muzzle, where a full facial hold was applied to suffocate it. The hindquarters have been totally consumed.

After placing the carcass in front of camera, usually between three and ten metres, nothing would usually happen until at least the next evening, regardless as to whether or not it was a new spot for the carcass or a regular place. First, it would be the magpies that would spot it. They would flit around; peck the ears, eyes or anus. Carrion crows would usually then turn up, followed by a raven or two. If the ravens found it early, and after smaller corvids had started to eat, they would eat. They would punch a hole in its ribcage and pluck out the lungs and heart if they could reach it. If it were a small deer, the chances were that they could.

Ravens are large birds, especially older females and they are very bright birds. They are cautious but not in the way in regard to the scent of humans. They certainly know if the bait has been placed, but it would seem that they do not approach the body first. They let the carrion crow and magpies eat or pick at it for a day or two. A buzzard may come in the next day and, at first, sums up the situation. At first, it may not feed but hang around on the body or nearby, checking the situation. If the buzzard feeds, then ravens will come. A raven rarely opened up a carcass without other corvids being at the scene first.

When ravens came, they did not feed for long periods of time or even hang around. The buzzards, on the other hand, were in control for long time periods, and hung around for over a week, often consuming the whole carcass. They may be on the body five times a day and remain there for over an hour at a sitting. Some buzzards started to feed long before dawn in the black of the night and leave at dawn. Although a buzzard is not particularly a big bird, a raven is bigger in some respects, but the raven will take food away to stash. The buzzard will not do this but just feed from the carcass, often until it has all gone.

When one finds a deer stripped of all flesh, it is easy to assume that a large mammal has been responsible, but in reality, it could have all been done by a bird or two. Birds do not eat large bones, but small ribs and leg bones, or skulls from rabbits and small hares or very small fox cubs. At a camera carcass, a buzzard ate a whole roe buck (except bones and most of the skin) not just once but three times, and different birds twice. When the buzzard feeds, it will take lots of skin and cut it quite neatly. If the deer is fresh and the weather is cool, and the bird needs a lot of roughage, it may take lots of skin and from the ribcage. In just several hours, a large area the size of the whole rib cage area of one side can be stripped. This may look odd and look like typical cat feeding traits. One must realize that fresh skin is stretched tightly across a ribcage naturally and when it is broken, it will recede, thus giving an impression that more skin has been removed than it actually has.

Cats feed on a lot of skin and so do foxes but rarely as much as cats do. It varies. For all the scavengers, it is the flesh that really counts and so more of that would be gone than skin. Usually, a cat eaten kill has skin and flesh taken from the same spot, especially on the rump, shoulder and bottom jaw, and ribcage. A cat will rarely eat the rump and leave the skin flopping around. That is vulture style eating. The cat, though will lick an area on the stomach region to soften it up and lick off the hair. It will then open it up, often removing the rumen first before opening up the ribcage by biting off one rib, then another until it can easily access the organs, so they are usually the first to be eaten. Then the cat may start on the hind legs and rump.

Other cats will move to the head region first and lick the head, and eat the tongue and skin around the face before eating the shoulder. It may then will skin the animal, especially if it feeds from the head first, and then usually pulling it back as it feeds so that eventually all that is left is the main back and some of the side skin with perhaps the hind leg bones attached. This is a typical way in which I have found large cats to eat most bovids.

The cat may need to sheer the skin away from the flesh as it moves down the body and, doing so, leaves striations on the inner side of the skin that looks similar to the sand on a damp beach when the tide has gone out. The carnassial teeth leave trenches of a certain diameter as it parts the skin flesh from the bulkier body flesh.

At times if a large cat is not very hungry or perhaps needs to leave its kill, it may just eat part of it and may eat some facial skin and flesh along with the neck and part of the thorax. A front leg may be missing and large areas of skin taken from the side. At other times the consumption may have started from the rear and in this case, the whole or part haunch may be taken. It has been suggested that leopards will disembowel all the animals they take before eating. Ideally, this may be true, but I think that most studies undertaken would have been on leopards in Africa and India, where the animals learn that it is in their best interests to do that.

I am sure that leopards also do this in Britain but maybe not as much. To leave the gut in within hot temperatures would lead the animal to bloat and become gassy, and smelly and attract more flies and thus more scavengers. Also, the gut contents would taint the taste of the meat. It is true that the herbage the deer eats is usually bitter but not always, and grass is not usually bitter, but other plants may be.

It is worthy to note that leopards do not do this with smaller mammals but eat the whole thing, as do most other cats and other carnivores, benefiting from the vegetable matter also. Cats will often leave the guts of rabbits or hares in a neat pile after consuming the whole the rest of the carcass. Although cats eat less vegetation than, say, wolves or foxes, they still need some and cats do know of natural plants that help ailments in just the same way as dogs do.

I have found grass in suspected large cat scats. Mustelids, on the other hand, eat even less or none at all, their digestive system is less likely to be able to digest plant cellulose, but a cat can do so in small amounts. It mainly depends on whether the cat needs the contents or not.

Many deer carcasses that I have come across certainly seem to have had the gut contents removed, often to the side. Other times, they are nowhere to be seen. It makes sense to assume that the cat needs to move the huge bulky stomachs aside so that it can gain easy access to the organs at the front and as the cats, more often than not, start to eat from the back and work their way forward, then removing the rumen makes way. Sometimes the contents have been clearly eaten and to back this up, I have found cat scats made up of the digested remains of rumen contents. Maybe the cats can deal with partially digested plants to obtain certain minerals or sugars that cannot be extracted from the flesh. Eating the partly digested vegetation could make sense. A cat, just like a dog, will pass through grass and not digest any of it. The grass is eaten to aid digestion if there is not enough roughage within the meal. It can also help clean the gut and boost the immune system. The gut of herbivores is heavy and another reason why large cats remove it would be to make the carcass lighter to drag away or to store up a tree if needed.

Most cats will eat the whole carcass or as much as possible, leaving the most difficult bones until last. Smaller bones will aid digestion and be either partially digested or fully digested. Most cats seem to eat at least some of the bones if their teeth are ok and they are not in pain from an abscess or other mouth or jaw problems. Some cats seem to eat far more bones than other members of the same species.

Cats seem to have at least a few individual traits within their feeding methods. The classic leopard eaten carcass would consist of mainly flesh taken but also lots of skin, one limb removed with no trace, with rumen removed, but with the majority of the skin mainly intact. The animal would normally eat the rump first and pull the skin over the head as it goes, often turning the animal inside out, and this can be seen if the animal has left the carcass for any reason. The ribs will be eaten down to the base or as much as possible. Removal of ribs allows access into the thoracic cavity to eat the more nutritious organs such as the heart and lungs. Often one side of the ribs are removed or just two or three, whatever enables the head or paw to enter. Sometimes they are cut in half, or eaten right down to the spine. Sometimes no ribs are removed, but organs are absent. At other times a rib or two are broken by the initial swipe of a paw, thus allowing organ removal later without actually biting off the rib. The ribs are the easiest bones to break off and eat and the cat often does this after eating the haunch.

The carnassial teeth of cats are basically sheering tools and are used like sheers to slice through flesh and bone. The carnassial teeth of leopards are big and range from one and a half to two and a half centimetres in length. They have immense strength and large, powerful facial muscles that both leopard and puma use to assert strength to easily break bones cleanly.

The ribs of deer can be cut across neatly and low down. If most of the ribs are cleanly cut, then it is most likely to have been a cat consuming it. Foxes have much weaker jaw muscles and lack the strength to break them. They are often gnawed down by foxes, or the small ribs may be broken cleanly by large foxes. The fox does not need to eat as much bone as the cat and as they are generally small animals, they concentrate on the easiest parts, the flesh. Foxes will eat bones if the flesh has gone and they are still hungry. The large cat, though, may decide to eat bone there and then and will enjoy a slow neat feed slowly manoeuvring around the carcass. If the cat is in no hurry and feels secure eating the animal, then it may follow a more usual pattern of feeding, as described earlier. If the cat is in a hurry, then it may just eat the easiest parts and then maybe tackle the more difficult and less important parts later. The cat may abandon the carcass after a single feed or return, depending on the security of the carcass and well being. Many times the cat doesn't get a chance to feed at all due to human or dog disturbance or even from noisy crows. Corvids can be so noisy and bothersome to large cats that they may abandon carcasses if they feel their cover may be blown.

The cat likes the head region and, if secure, will eat the muzzle or the throat, including the tongue and areas of skin that are usually neatly sheered, causing a circular shaped hole in the pelt as the skin retracts from its stretch. A cat usually eats in a neat way, not making much mess. The cat is a clean animal and it shows within its eating habits. Most cat species are solitary living, unlike a pride of lions, for example, that may have competition and so not eat in a slow and finicky manner. Dogs, on the other hand, like wolves or a pack of domestic dogs, compete and so the carcass will be ravaged, broken, dismembered and generally messy, although these animals can eat quite cleanly but still not as with the neatness, of which a cat deploys. A single dog can make light work of a deer carcass and there can be many similarities, especially regarding the consumption of bone and cleanly cut ribs. One must imagine how the cat feeds and look at it from the cat's perspective. The cat thinks before taking action, this is within the general cat mode, not just for feeding. So when we find a carcass, firstly, one has to assess whether or not the animal was killed by a cat. There are many points to search for regarding whether or not a deer, for example, was not killed on a road or by a bullet, asphyxiation, or by exhaustion or by any other means. All things must be taken into account.

The unknown factor in this field is the fact that we are not one hundred per cent sure about the species of large cats responsible for killing prey, and although many species have their own particular traits, they are all individuals and often do things differently. It would seem that many animal carcasses have the same kind of injuries on them. Most large cats kill their prey in three main ways. By biting the back of the head or neck and severing the spinal cord, or secondly biting and clamping onto the underside of the neck to squash the windpipe to suffocate, or a full facial hold to suffocate. The windpipe clamping method is the preferred way in which the leopard deploys its tactics and second to that, it will bite the back of the neck. Most cat species will bite the neck; even small cats bite the necks of small mammals or, if they can, bite through the skull. Large cats cannot do this easily but have to find a way of immobilizing the animal as quickly as possible whilst at the same time causing it to pass out as quickly as possible, which means either stopping the heart or preventing oxygen from entering the lungs, and so the brain. One of the quickest ways of doing this is by suffocating the prey. Usually, a bite to the throat is the easiest way of administering asphyxiation as the cat can hold the animal down securely under its belly. Both leopard and puma have been seen to hold the face, and there is film footage and still photographs of this happening.

In some parts of America, it is well known that the puma bites the nose, often right off, if the animal struggles much. Why then am I finding so many carcasses with wounds to the face? All across Britain, carcasses are being found with their noses bitten off and not just sheep and deer but badgers also and foxes. Many seem to have this hallmark! On many of the animals, they have evidence of neck injuries inside, and one can see where the cat had gripped the neck due to bruising inside. The neck skin is not often pierced by the canine teeth, so one has to open up the neck to look.

Many cats do like to eat the soft, moist nose, and also ears and other external bits, but on many of the animals, there is no evidence of normal neck biting. This surely means that the predator was intent on killing by giving a full facial hold or at least biting onto the nose (this may cause a stunning sensation to the prey and so easier to control). In many cases, the nose will be bitten off after the animal has been killed, but in many, it would seem that the biting on the nose or muzzle was the way in which the animal died.

Lions use this method more often than not to suffocate big animals, and it is easier as there are many lions to hold down the prey whilst another concentrates on killing it using this method. The cats in Britain are surely single animals, not prides of lions, unless some of the cats go around in pairs or more. Usually, large cats such as pumas and leopards do not hunt together as a team, so why are these findings becoming commonplace? One idea is that we all have it wrong and that it is simply dogs that are fooling us all. I can imagine pit bull terriers or Staffordshire terriers doing this and it may be that some dog owners teach their dogs to hold on by the nose despite the animal's struggles. Many dogs have such stamina that they could do this. But would the dogs then eat some of the dead animals, in a clean way, and drag it under a bush then disappear without returning to feed on the rest? And why had not the owner taken much of the meat that remained, even for his dog? It does not make sense. But if we had some weird kind of hybrid involved in the large cats, then that may explain a few things.

The animal carcasses that are found with the nose bitten off or severely damaged are often out of the way, or in people's gardens, or under a bush by a river, or even within fenced areas where dogs do not go or even can't get in due to fences. We do have a strange dilemma on our hands. I do not think that dogs are responsible for the majority of these cases, but I think leopards or pumas are the culprits. There is much disbelief that leopards kill in this way, but I have seen an old wildlife documentary clearly showing a leopard killing a wildebeest in this way. They do deploy this method, although there is not much documentation on it, I have seen other more recently made wildlife documentaries, where a leopard held down the struggling animal by biting the face and the animal quickly submits, and pumas are known to do it to cattle and horses.

Occasionally, I will find a deer carcass with broken ribs, usually on the right side. I was at first puzzled by this, but later, I read that pumas often do this to bring down animals. Maybe a broken rib could slice through the lungs, heart, or adjoining arteries. A deer could be winded and so not be able to move far. It is hard to accept that a large cat knows this. But should it? Cats are capable of knowing about bodies and organs and what their function is, they must do to some extent, although most of it is more likely just instinct rather than forethought. I am sure that cats break the ribs to gain access to the main organs within. But many an older cat may learn that a swipe to the ribs would render the struggling animal helpless and disable it enough to make the kill. At one of my study sites, sixty per cent of the sika hinds that were killed had broken ribs seemingly before the removal of flesh and one rib had gone through a lung.

The remains of a sika hind near Bere Regis. This sika hind was taken in a strip of woodland. The body was dragged under a fallen pine tree and consumed maybe by more than one animal. The skin has been removed and the limbs are missing. The head has been neatly sheared. On the snout, there are the marks where incisor teeth have gripped onto it. Puma regularly brings the animal down by a bite to the nose and leopards do the same but are usually adept at a full facial hold. The method more often used by both species is a bite to the neck underneath or dorsally.

This is the other side of the head, showing the neat cropping of flesh and bone and skin that is so typical of cats feeding habits. Foxes will also do this to some extent, but it is overly less neat and they concentrate on bulky areas of flesh if it is available, breaking amounts off and carrying it away to bury before coming back for more.

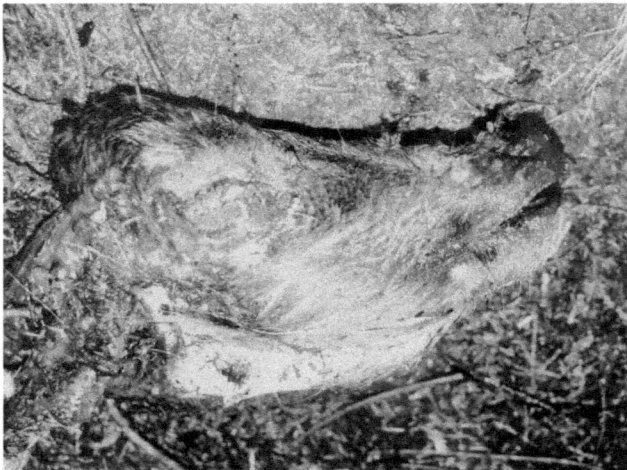

A close up on the head showing canine hold marks.

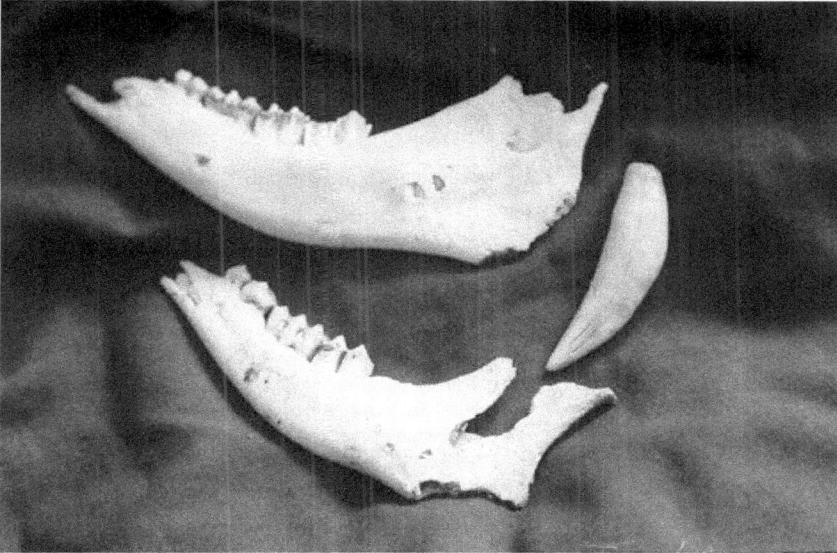

This is a lower jaw bone of a sika deer and has a larger slice taken out of the side and the leopard tooth also fits this in dimension.

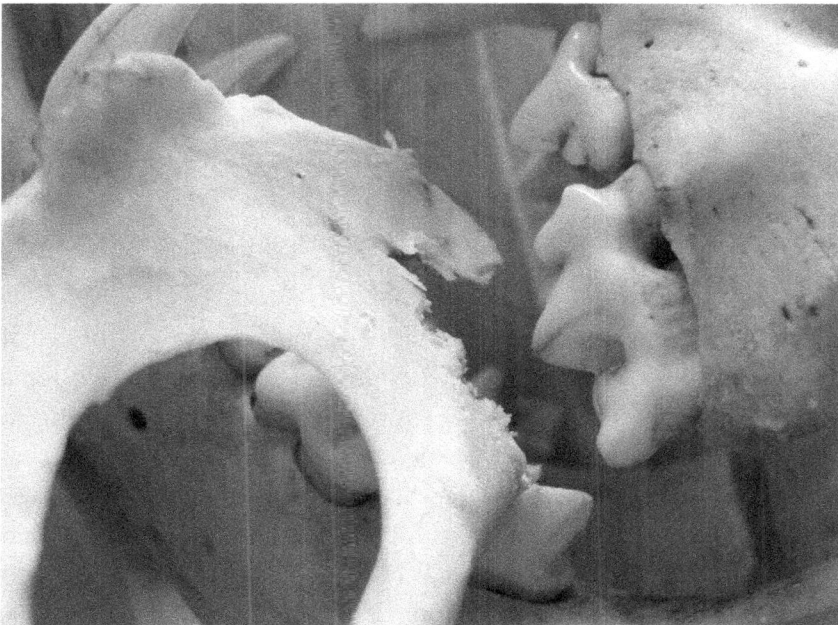

The pelvis is the most likely bone to find Tooth pits and carnassial edging on deer bones.

The bitten nose may be due to the cat trying to drag the animal, but that is very uncat-like, as a leopard will straddle a carcass and hold it firmly on the neck or skin around the front of the body, and the animal slides along with legs folded back for easy manoeuvring. But for a cat to grab the muzzle and drag the animal backwards, it is not easy and cats do things in the easiest and least hard ways.

Usually, the nose is not bitten off but damaged due to the sharp incisor teeth slicing into the soft nose or surrounding skin. Incisor teeth are often the sharpest of teeth that the cat has within its mouth and they are like chisels. With pressure they slice through flesh.

Often one can see where the canines have held into the sides and they do not penetrate as the jaw skin is very tough and stretchy, so there are usually small bare areas of fur where the canine tips have rubbed away the fur. A cat that has sharp canine teeth will often pierce the neck skin and when it does, the teeth get to work immediately by breaking the windpipe and causing much bleeding and bruising.

For the cat to quickly kill the animal, the jaws have to be clenched very tight, so the teeth will meet on a small necked animal. There are rarely teeth marks on the neck vertebrae. Often the neck areas of prey are eaten out, so it is impossible to find the cause of death, and many animals are scavenged after natural or unnatural death by large cats and then consumed in a normal way.

There is certainly much hunting with dogs and many types of use of various hunting and running or fighting dogs are being used for a huge number of very cruel blood sports. Dogfighting is gaining in popularity and making dogs stronger and braver, the nasty owners are using badgers, foxes and deer as bait to toughen up their dogs. This will obviously have a confusing effect on the large cat investigations. Only large cats will bury a carcass under the ground litter such as leaves or sticks. Pumas and leopards often do this. Storing carcasses in trees is less common in Britain, but remains have been found many times, especially deer remains high up. There may have been one or two hoaxes also. The leopard is the cat known to habitually store prey in trees and it is mainly to prevent other scavengers from finding it, such as hyenas, jackals, vultures and other large cats. It is partially instinctive but in areas where there are not so many threats, the leopard may not bother so much. Any large prey remains found in trees could be attributed to leopard activity if human intervention can be ruled out. Leopards may stash any large prey in trees, including swans, dogs, geese, hares, badgers and foxes.

The canine teeth of large cats are used as holding tools when manoeuvring around the carcass and many canine tooth pits are to be found often at the ends of bones in the joint sockets or balls. Often, the base of the bottom jaws is often bitten off cleanly. The skull often has the brain cavity eaten away, often partly or the whole of it eaten, including large thick-skulled animals. Often the carcass may have been dragged some distance.

This is also not proof that a large cat has been responsible as foxes and occasionally badgers will accidentally move it by trying to tuck flesh from it, usually when the carcass has been devoured of more accessible flesh. The tugging fox could move a large carcass over several hours or in a night, for example, several meters or even up to twenty meters if it is in an open area such as a field of grass. This also confuses the large cat detective.

When a puma or leopard moves a kill, it holds the carcass high off the ground to reduce drag, and doing so will often hold the animal clear off the ground if it is smaller than itself and if it is larger, then one may only see the drag marks made by the hind legs or the rump. Still a large drag mark will be seen in soft ground or in vegetation. There have been many cases where farm animals such as sheep have been dragged and the cat has leapt with it over a fence or ditch. Such is the strength of these animals.

Leopards will drag a carcass a long way if they feel the need to and will readily take them over five hundred yards to safety. If the cat feels secure, it may eat there and then where the animal was captured.

Sometimes the leopard will drag the animal alive whilst throttling it to get away from an uncomfortable situation, even up a tree if it has to. A leopard may consume a deer at the edge of a road, even if there is much traffic about if it feels safe to do so. If the cat feels unsafe, then it may abandon its kill and either return later or not risk the chance of harm by humans or other animals. Most cats will prefer to eat at leisure and, if not at the place of the kill, then elsewhere. It is at these places where one can see the typical eating habits of large cats, but before the other scavengers get there. One can often see a depression in the ground around the carcass where the animal had sat and fed. Sometimes two or three cubs will eat alongside flattening down more grass. Usually, the cat will not move around to different parts of the carcass but move the carcass towards itself. Often the remains are turned upside down or at an odd angle, sometimes balancing in mid-air. A strong puma or leopard can manoeuvre the heaviest of deer into the right position

The skin is often removed, and often the liver and lungs are usually eaten along with the heart.

The rumen is usually removed, taken out of the body cavity and left to the side. Large cats may or may not eat some of the stomach contents. I have found ruminated plant remains within some scats. Foxes may take the rumen if it is a few metres away from the main carcass as they usually are very reluctant to tuck in, often for days afterwards of a large cat feeding.

Fresh carcasses can be more difficult to assess if not much flesh is removed, but a few key points can be apparent.

Often a musky smell may be present, similar to a wet dog, or may smell like a dog eaten bone or another item. The smell is made in just the same way but does not retain a doglike smell but more of a wild animal smell, but not too distant from the wet dog smell. If dogs have been consuming the carcass, then it will smell similar but more to a typical dog. The deer may smell of deer scent, which is musky, but it has an earthly warm musky smell, especially if it is male and even more so if the deer was killed during its rutting season and the animal was rutting. This may be too strong and cover up any possible cat smells.

Cats usually remove much skin. They have the sheering carnassial teeth to tear through large amounts and as they don't chew, they just swallow large chunks. This often comes out in their scats as tightly twisted rope-like sections. Large circular areas of skin may be removed from the face, shoulder or sides. This is often apparent on large animals; smaller animals are eaten in a slightly different way, as to leave nothing but perhaps a fragment of head or legs or pelvis.

Ears and nose are usually consumed and more likely to be so if cubs are present with the parent. The skin edges may be licked up and if one looks carefully, one can see where the tongue has been at work, smoothing out the skin and creating points at the end. The cat's tongue is like sandpaper to some extent. It has fine tubercles covering its surface purposefully there to rasp away at flesh. When one considers the cause and effect, one can see it quite plainly, if, indeed, there is licked up the skin to search for it in the first place. The tongue also gets to work on the bones and they are often licked clean.

If compared to a carcass in which foxes or dogs have been eating, it can be quite apparent that the latter do not fuss about like the cats when feeding and rarely bother to lick areas clean. There is some tongue work, especially with dogs, if they are allowed time at a carcass, but this is rarely the case and one never finds the same hallmarks as one does with a cat eaten carcass. There is rarely much mess around the initial site of first consumption. If the animal was consumed soon after death, then blood would have messed the ground but not in a big way because after the animal suffers a heart attack, the blood often clots at the heart or in the neck or abdominal cavity and so it all congeals and becomes concentrated, thus being easy to find or see unless a person does an autopsy by opening up the whole of the inside of the carcass. It may look like the animal was sucked of its blood, as many people will say. Much blood remains within the main muscles and is eaten along with the flesh.

Plucking is often seen and large amounts of fur will be around the initial site where the cat has plucked it out with its incisor teeth, often leaving rings of fur around the rear end or belly, or groin. Sometimes much hair is plucked out, at other times, only a small amount and other times, no plucked fur can be found. If a deer is moulting, then lots of clumps of hair will be naturally falling out and can give a false impression of plucking more than it actually has.

The legs are usually pulled inside out to reach the lower muscles at the ankles as the cat works its way down each leg. Often only one leg will be turned inside out to the hoof. Only a very strong animal can do this and usually, a human cannot if he or she tries.

Analysing The Tooth Pitting And Trauma

Most feeding will leave some sort of marking on bones, especially if the felid consumes most of the flesh from the bones. Most of the bones with tooth trauma on them are swallowed, thus not allowing it to be found (except some from scat remains).

The most common trauma to be found is indentations from the canine teeth. Secondly, the carnassial teeth when sheering through bones, or by accident. Thirdly by incisors.

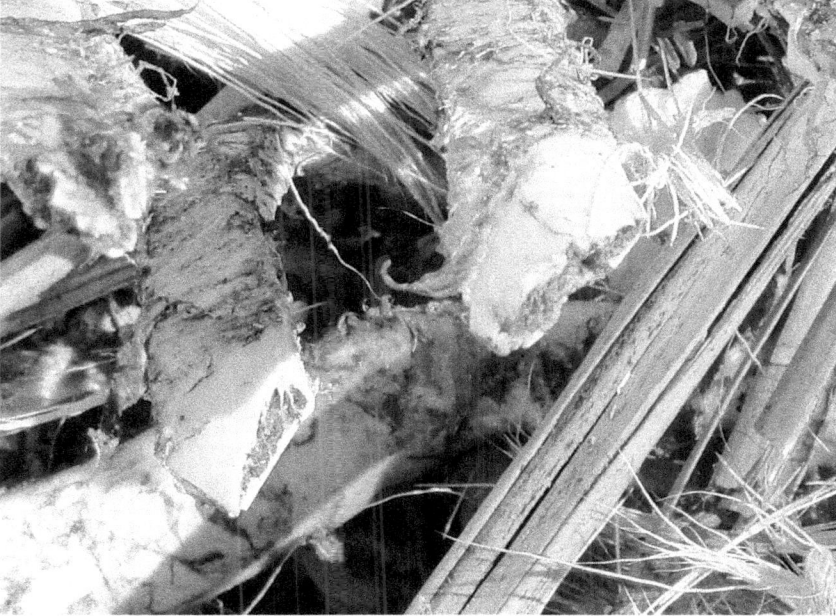

The carnassial teeth of leopards and pumas are very strong and are able to sheer through bones cleanly with a lot of muscle pressure behind them. All species of cats eat large amounts of bone. The leopard may prefer to consume larger bones than most pumas and leopard kills often have the larger bones consumed or partly so. A fox could not shear through these bones as cleanly as this. Badgers rarely eat carrion, especially if their preferred food, earthworms and insect larvae, are abundant, but if they do, then they rarely eat the large bones. There will always be one individual badger within a colony that is partial to fresh carrion occasionally, but even then, they are not clean eaters and they find it difficult to take the flesh off the bone, dragging the carcass around as they try to get a grip on things. The teeth of badgers are strong but small, and the carnassial teeth are not for sheering through bone but for crushing roots and bulbs and nuts, similar to pigs. Badgers are even warier of dead deer and livestock than foxes. In many tests I have done regarding the carrion-eating habits of badgers, most will not touch carrion until the maggots have got big and then eat them when they have been washed by rain. In the winter, if the ground is cold or frozen and earthworms are driven far below ground level, then badgers are more likely to take advantage of such meals. These ribs are from a deer carcass that had all the hallmarks of being killed and consumed by a large cat.

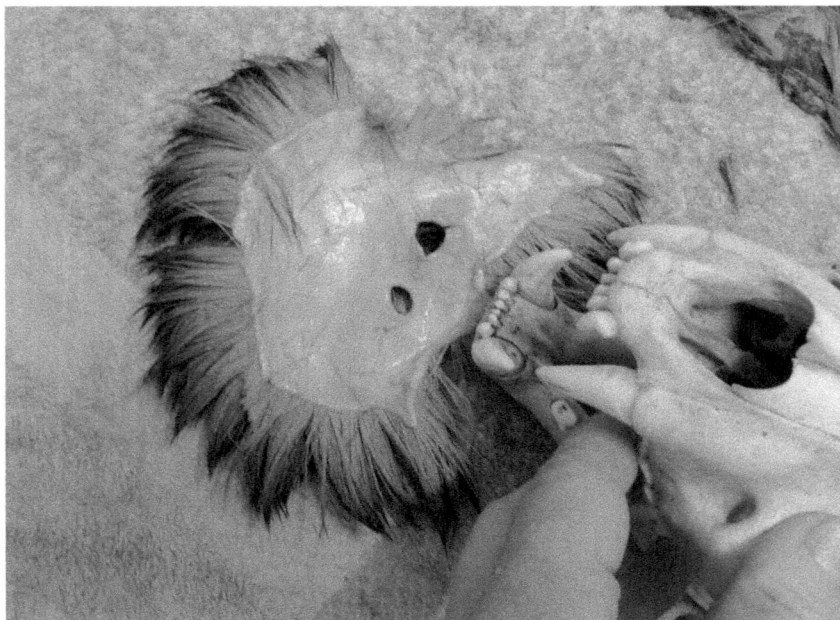

Another piece of skin with the canine tooth holes with a leopard skull for comparison.

The impact from canines can be found on leg bones or in the skull, jaw or vertebrae. Often nit is a hole punched right through the bone rather than just a tip imprint. Cats often hold and bite through a bone with the upper and lower canines, usually a leg bone. Holes can be found in scapulas and sometimes on the ends of the pelvic girdle. Most canine trauma is evident on the hide, especially the neck or head areas.

The main and possible most important of tooth pitting comes from the carnassial teeth. Cats have far fewer teeth than canids, as they do not chew as many dogs do; the molars have evolved into slicing rather than crushing teeth. In the top jaw, a cat only has two on each side, but the hind ones are like two or three teeth joined together. They form a line of cones along with a side cone that is lower down and rarely makes contact with bone. The three cusps of these teeth have what are known as Paracone, Parastyle and Protocone.

Whist the paracone and parastyle often make contact, the protocone being on the inner side and lower down, rarely if ever makes contact unless a bone is angled in such a way with irregular contours. The bone needs to have higher areas for it to make contact, and even then it is usually slight and rarely deep, whereas the other two cones regularly make deep or light impact. The chances of all three cones touching bone are rare. The paracone is more obvious in young animals and seems to wear down as the animal and teeth grow. However, if all three cones touch down, then an imprint can be formed that looks like an irregular triangle of small pits. These are called cone triangles. They have a consistent arrangement of which is only typical of felids and not canids, and the distancing between each cone could suggest the overall size of the individual animal and species. The cusps of a dog will be smaller in general, even on a large dog. Thus, one can assume that a large cat rather than a dog was responsible. A large dog may have approximately three millimetres in-between the cones at certain points, whereas a cat maybe five, six, or seven.

The places most likely to be found are the pelvic girdle; the reason being is that when a cat ploughs into the haunch, it hits the pelvic girdles on either side and as the cat manoeuvres around it, or onto it, then it is these teeth that may hit it if it is at a certain angle. The cat tests the bone with its teeth and decides whether or not to crunch into it. Usually, it does to some extent, often breaking of the ends and consuming them, or if the individual decides that it is too hard will leave it after having made those test imprints. This is when the cone imprints can be found, but more often than not only two will make contact, the paracone and parastyle. The protocone may make a slight mark or not depending on how much it has worn down because the cone wears down suggests that at first, the cone must be longer and has to hit bone many times for this to happen. So one can conclude that is mainly young animals that can cause such trauma. This in itself causes a discrepancy! Because if it is a small young cat, the distancing will be small, causing the interpreter to assume dog rather than cat. Thus, many triangles may be written off as being from canid rather than felid.

This is not an exact science, but it is very helpful, and at least one university is using this analogy to ascertain whether or not a large cat has fed from a carcass. It is slightly flawed, but it is also a great tool.

I find all three cone impacts rarely, and those that I find are gold dust and are saved and passed on to the agricultural university in Cirencester. Most are on the pelvic girdles, rarely scapulas, and occasionally on limb bones. The best media to find them on are the growing antlers of male deer or horns of sheep or goats. The growing antlers of deer are soft and full of nutrients and many cats will eat them down if still small leaving great triangles on them. I have only twice found triangles on hard antlers, usually from curious cubs.

Triangles can also be found on supple skin, especially around the neck or head.

Typical sheering can show up the exact dimensions of all the other few teeth on any bones but mainly the pelvic girdle again. One can also note how worn the teeth are, so giving the possible age of individuals.

Often, the cat will eat one or two large legs of a deer completely; larger deer and farm animals may have stronger bones and be left or carried away by foxes. Badgers never carry food away or cache food but usually eat it there and then on the spot, and usually, badgers do not touch carcasses, especially if a large cat has been eating it. Badgers usually eat deer carcasses if the weather is very hot or very cold, preventing the usual consumption of earthworms. Foxes are the main culprits of scavenging after a cat has finished eating, and secondary will be buzzards in areas with high numbers of these birds or, thirdly, corvids, crows and magpies.

The older carcass will present tooth pits more clearly, as all the flesh has been removed by drying out, slugs and other invertebrates consuming minuscule veins and skin particles, then pitting will be revealed. Now one can really see the trauma caused by the teeth of the cat. The canine teeth break through the bone ends, pelvis and femurs at the ends, creating typical large rounded holes, sometimes for several millimetres, other times several cm or over an inch in some joints. The whole length of the canine can be imprinted into the bones of the jaw or ends or joints on the tibia or femur. Scapulas also can have neat imprints. Whilst the carnassial teeth are sheer, they do not always do their job on tougher bones and often, the cat will test or try on a bone only to realise that they are too tough and abandon that notion.

(See photos on cat kills).

The royal agricultural college at Cirencester has taken a leading role in analysing tooth pits over recent years. Andrew Hemmings is initializing a tooth pit project with students and so far has made good progress, with several big cat tooth pits being scientifically identified from bones submitted by myself and other large cat researchers. Not a lot had been done in the past by scientific establishments regarding the evidence of large cats in Britain, so when the university offered to help, it was about time that an institution took it seriously and wanted to help submit the scientific evidence. Without this, we would not progress.

Most of the bones with imprints submitted by myself were from Sika deer initially, but I also submitted badger bones.

One can look at twenty deer carcasses and not find a clear triangle, but this is the evidence used to determine whether a canine or felid had been at work on a carcass. I think it worthwhile to look at all of the tooth imprints and use them in conjunction with one another using a template. I have made up small clay tablets with tooth impressions pressed into them as a guide when out in the field or in the lab. This is very helpful as it picks up on the smallest of marks and they too can be measured in relation to the real dimensions of a leopard and puma skull.

There can be many problems when trying to assess what animal was responsible for the consumption of the flesh from any carcass. As I have stated before, it is rarely one single species responsible unless a carcass has been found buried under vegetation. Pumas and leopard hide their kills under bushes or rock crevices if available. If not, surrounding debris will be pawed over the carcass concealing it from sight. This may also help mask some of the smell, but mainly from sight.

Animals with colour vision can see red and bloody muscle and bone can be seen a long way off in a green or brown environment. Birds generally have exceptional colour vision and can give away a carcass to other predators. Vultures, for example, are clued up to see such things as well as smell them.

I recall the time when I found a roe carcass buried under pine needles by a puma, I could not see it but smell it and it was only when I was on top of it that the antlers gave it away as they protruded from the debris. Often the carcass will be well covered with sticks and leaves and it may look as though a human has done it. When such covered up carcasses are found, they are rarely whole, but some consumption would most likely to have taken place. The rump or shoulder may have gone, for example, or in the least, the telltale puncture wounds of canine teeth will be on the back or the front of the neck or throat region. The nose may be damaged or gone completely. There may be territorial scats placed nearby, especially in the case of puma kills. Cat cubs also cannot control their bowel movements as much as adults and where cubs are concerned, there may be small scats at the carcass site or even on the carcass.

Foxes do this also to try and take claim of a carcass from either cats or other foxes, usually other foxes. A cat may not return to a kill after foxes have mauled it. Foxes cannot bury carrion with debris, but they cache items, usually burying them in the ground. Foxes can clean a carcass by biting off large chunks of flesh, burying them and returning for more. They are basically filling their larders for future days or even weeks to come. Sometimes one can come across a carcass that has been stripped of flesh within a night, and it is so easy to just assume that a large cat must have taken all that flesh, but in reality, a fox may have been busy going to and frow all night! It is then that all the other typical evidence comes into play. Sometimes there are no typical hallmarks of a large cat on a carcass, whatever the decomposition or whatever amount of flesh, bone or skin taken, usually though there are. A carcass that still has much flesh on it may not show up with tooth pits, so best to wait until the bones are clean. The best thing to do is either place the bones near a nest of wood ants or put it securely in a tree. This way, no foxes or dogs can move the remains and they will eventually become clean.

When a cat eats flesh from bone, although it will sheer off large chunks, when it comes to bone, it will lick, and then the blood veins, which are harder than the surrounding softer flesh remain and show a licked up state, often the spine and leg bones will show this, and it concluded that only a cat was responsible.

Skin is also licked up, especially on the head and cheek areas. Canids rarely have the time to do this.

Dogs At Carcasses

There is much confusion about the ways in which cats and dogs eat. Many people do not know how to tell the difference, of which there are many. Some people claim that there is no way one can differentiate between animal bodies that have been eaten by either animal; this is not true for many reasons. People drive out from town and cities or neighbouring villages and just let their dogs out to roam and cause havoc until they return home. These people are oblivious as to what their dogs have been up to to a large extent, or they do not want to know, just like a person who takes a dog for a walk, knowing that it is going to pass its waste before long, so walk ahead and do not look back in ignorance. This is bad enough, but to just let the dogs go off and do what they want is clear irresponsibility, yet many people do it all the time. I see it all the time and when I used to spend a lot of time out photographing wildlife, I was constantly seeing dogs roaming without their owners. Many people spend hours calling their dogs after they have run off chasing deer, hares or other animals. Many dogs go back to their owners with blood on their faces. This blood is either from where they have bitten their own tongues or from the blood of deer or sheep that they have harmed.

The issues with dogs are detailed, and I could write a whole book on the subject. I love dogs but dislike many of the humans who breed and keep them. There are way too many dog owners all over the world, let alone in the UK. Dogs have just become fashion accessories rather than kept for real need. People take it for granted and it seems that dogs are part of our cultural past and present, and unfortunately, the future. We do not need dogs now; they were used to aid hunting when we were hunters and gatherers and helped us climb the ladder of evolutionary change to a small extent. When I state need, I mean just that. There is a difference between need and want. The high majority of dog owners want, not need. Dogs are really wolves and belong in the wild, not cooped up in hot cars, kitchens covered in poisonous cleaning agents and chemicals or on ladies' laps. Modern living is totally against the dog's natural way. Most dogs do not live fulfilling, happy lives but are forced to live in unnatural ways like humans. They are not people but animals with animal needs.

People do not think or really care about the animals they keep and with limited intelligence, most dogs are tortured to some extent. One of the reasons why dogs are badly behaved is simply because of humans and the way in which dogs are not allowed to live out their natural abilities. The dogs are doing what comes naturally to them. Of course, there are many well-behaved dogs and good owners, but that does not mean that it is acceptable in my eyes. Humans keep dogs for selfish reasons, and I basically disagree with it because we are living in different times to that of a few hundred years ago.

Many people do not know enough about dogs, and the breeds and what they were made for. Dogs were all bred and genetically modified for hunting purposes. We do not need to hunt anymore, so most dog breeds are now invalid, yet they retain their given abilities, much to the ignorance of their keepers a lot of the time. Obviously, there are many hunting types of people who are not spiritually aware and think that we are supposed to be living like ancient man when we should have moved on.

Many people do not associate dogs with wolves or other carnivores because of conditioning, so many will not know what their own animal is capable of. So many people claim that their dogs will not chase sheep, but it has been proven over and over again that many do and often behind their owners' prying eyes. The vast majority of sheep deaths in the UK can be attributed to dogs.

Many are from cats and some are from other predators. Large exotic cats are not the only predator to be now naturalising Britain, but it may be that some of our other natural apex predators also exist, alongside other non-natives. Again I use the term non-native loosely as all things are native in the long term, past or future tense.

Brown bears, Grey wolves, and wolverines exist in small numbers, and other predators such as jackals and non-native fox species occur in small numbers but not on a naturalised level. Wolves have been known to live in Britain for the whole of their lives. They must have left typical field signs all over the place but perhaps kept a low profile. A lone wolf does not advertise its presence, and a pair will also be very quiet.

A pair lived in Northumberland for at least nine years, unrecorded by most people. Sheep that were eaten were thought to be the work of stray dogs, occasionally large cats. Because we were not told the truth about these wolves, or the many wolverines that were living in the West Country, then people would have made many mistakes.

Wolverines are predators, unlike their social relatives, the common badger. Domestic dogs are to blame for many multiple sheep deaths or injuries, whilst the disappearances of one or two ewes are more likely to be the work of a large cat.

Dogs usually do not target just one animal but usually chop and change, harassing many individuals and when one is dead, the chase is no longer happening, so it will harry another. This is basically the way in which a dog or a pack of dogs behaves. The dog is not really hunting, and certainly not for food, but for natural instinctive pleasure. Dogs need some pleasure in life, and what more does a dog need than to run and chase that is their natural given reason for living. Some dogs feel the need to exercise those traits and they cannot be blamed for doing so, but the owners who allow them to do it. The dog owners should give the dogs enough stimulants in other ways to discourage the natural trait from surfacing.

Large cats only kill to survive whilst dogs do so for entertainment mainly. Dogs that do attack sheep usually do so in typical canine ways. They attack at the shoulder or the anus usually but not always. Some dogs were bred to bring down wild animals and also to kill with bites to the head or neck. This trait does occur naturally in wolves but is not the main way of killing large animals.

Grey Wolves evolved alongside other large herbivores such as bison, and other bovids, including natural cattle-like beasts, Red deer or Elk. These large animals could not be brought down in the same ways that individual large cats can do. One individual wolf is not like one individual leopard or puma. The wolf has evolved to act as a team and all or many co-operate to bring large game down. A large bison, for example, can feed a large wolf pack. It makes sense to target one large animal rather than dozens of smaller individuals. The only problem is the sheer size of the animal's head and neck and so the wolf must find another way of immobilising the prey. The surest way to immobilize a bison is to injure it enough to cause later death from blood loss. By biting into arteries and pulling out the bowel, the animal is destined to succumb to whatever else happens to it. Thus, wolves usually do this on large animals rather than bite into the neck or head.

Most large herbivores sport headgear and so there is a great possibility of injury due to horns and antlers. Wolves have not the power of large cats. Wolves are running animals designed to chase their quarry over large distances, that is the best way a wolf and prey co-exist and the best way in which to wean out the weakest individual from a herd. The wolves will not attack the healthiest for obvious reasons. Nature does not work like that though there are exceptions to the rule.

Dogs, like their wolf descendants, are animals of stamina and endurance, as these traits were not necessarily taken out of the equation. Cats, on the other hand, are not long-distance running animals but stalking and ambush killers. A lone wolf or even two, can and often stalk and ambush, but it is not their preferred method. Also, a large cat can run over a certain distance to bring down an animal, but it is not its preferred method. When assessing a carcass, one must always bear these traits in mind, and in doing so, answers come naturally.

Dogs, whether one individual or a pack, will rarely take away the carcass. A single dog may find some carrion, such as a deer's leg or sheep head, for example and run off with it to its owner or some other bone. Rarely is the dog serious about it because it should not be hungry enough to do so, as most domestic dogs are fed at least once a day by their owners. This alone should be enough to deter any dog from hunting, but it does not work like that. We are talking millions of years of natural instinct. If a dog is very hungry, the natural instinct will materialise even more, but it is usually there in the first place when the dog registers movement, which triggers the chase.

This is, of course, the same with cats and even more so as they depend on movement more than anything else as a trigger. Dogs are more guided by scent, more so than their wolf cousins, as smelling abilities were selected and so dogs are genetically modified to have a very good sense of smell. Dogs working as a pair or more are more likely to multiple kills or maim as they get so hyped up over what they are doing. The adrenalin kicks in and they are off, one after another, not to eat but to chase, rip, and get a thrill. Many a pet dog goes out at night to get that thrill; the owners are not aware of what is going on.

In one respect, one could argue that the domestic dog is far more dangerous than a large cat, and to be blunt, that is correct. Perhaps when the 1976 legislation was drawn up, domestic dogs should have been added as dangerous animals. I put things into perspective in several places within this book, so there is no need for it here. Some dogs will eat from a sheep carcass if they are stray or are simply not kept well and undernourished. Stray dogs do not last long in Britain. Most domestic dogs here are kept indoors, with a family of people, so know nothing else. Most dogs would not have a clue how to survive out in the wild countryside and would soon get ill and perish unless they catch on to being a wolf again. This has been greatly diluted in many dog species and usually, any natural instinct is more tipped towards the most natural-looking breeds and especially those attributed to hunting.

That does not mean that small dogs never have the instinct because they often do. Small dogs with short legs were bred to enter burrows of mammals and so often still retain hunting instinct, especially regarding terriers. Sometimes these smaller dogs, including jack russell, highland, and border terriers, will try to chase but won't be able to keep up, but still frighten the daylights out of the sheep and even cause injury through panic and lameness. Often a terrier will catch a ewe and tug, causing much damage. They will easily kill lambs. The fighting dogs will also chase and when they capture, they kill and maim in nasty ways.

Still, it is the larger dog breeds that maim sheep, the most often culprit being the sheepdog itself! Yes, collies are the main culprits for sheep attacks; even a working individual may go off behind the owner's back after the normal working of the sheep to cause injury and to kill. The larger the dog, the greater damage can be done. If dogs do kill or sometimes maim and start to eat, they nearly always do so from behind, tugging out whatever they can from the anus, biting and making the hole larger to gain entry. If the bowel or gut is in any way protruding from the anus, then it is most likely to be the work of dogs. Often large amounts are pulled out before death or even without dying as in some cases. The dog may give up before more harm is done, or the owner may call it back.

Biting at the neck is often the first line of attack by a dog. Small sheep or lambs may be grounded by this method. Some larger dogs may bite the head or neck but rarely have the strength to cause the same amount of trauma that a large cat can. A large cat will firstly opt for the neck and is the most likely place to have trauma on. Large cats never go for the behind, it is strictly a dog thing, but we should be cautious in certain cases, and one is a parent cat with one or more cubs.

Sometimes when an adult is suffocating the animal, the cub may come in from behind and start to eat or try to drag down and leave tooth imprints. There will always be claw marks on most large cat kills and can be found if the hide is intact. Often the claw penetrates into the muscle, but this is rarely visible, as these parts are often eaten first.

Dogs usually eat from the most accessible areas, which, of course, is the anus. They rarely open up the animal from the belly, as most large cats will do. Most adult large cats will remove the rumen, which are always full of fodder, by raking it right out or to one side to avoid contamination of the flesh. Deer and sheep eat bitter leaves, which penetrate soft muscle tissue and render it useless. Also, there could be harmful bacteria or diseases within the gut and they may also know this. Sometimes a large cat will eat the gut, usually if there is little or no food within it, or even do what dogs do and that is to eat it, but in a way that pushes out the contents. Occasionally a large cat will eat the whole gut of smaller animals just to obtain the moisture or minerals needed for procreation. Large cats will often eat the growing antlers of deer and horn keratin for the same reason. Dogs are not fussed about the large guts of deer or sheep and will just rip it about without much care. Where a dog rips at the anus, tooth marks in the flesh are usually obvious. They rip and tear, they do not clean cut through flesh, like large cats do.

It is rare for dogs to consume a sheep or a deer carcass after killing it. It does happen, and the dogs are often owned by travellers or gipsies. When they do, they usually start from the anus consuming lower innards and the haunch, rarely consuming ribs. They may not eat any bone on a large animal and just consume the softer parts. Bones and skin start to disappear later when most of the available flesh has gone.

If more than one dog has been at the scene, the carcass, may be jerked this way and that as each dog usually tugs in different directions. There is always some competition between dogs and wolves regarding eating and it usually still remains in the way in which domestic dogs eat. The chance of domestic dogs having the time to consume a whole carcass is rare because it would usually involve several visits. This can happen.

I once found a field near Poole harbour that had been visited by at least one wild boar. It had ploughed up a large area of a corner of a field in the shade. I wanted to capture one on camera, so I placed a trail cam on a convenient overhanging willow and placed a road-killed roe buck under it. The boar didn't come anywhere near it, but a golden Labrador appeared every day around the same time to polish off the carcass, bones and all at the end. I later realized that the dog lived in the nearest house a stones through away.

I have also had hunting type dogs at many a carcass under cameras, but they have never even attempted to eat. The amount of dogs that have sniffed out deer carcasses at cameras and not touched them is too many to record. All I can say on the matter is, is that only twice have I had dogs eat a deer carcass put out for cat bait. Dogs that have lived a whole life eating wild animals may get the knack on how to consume a carcass and so become neater at doing it than, say, a young inexperienced dog. If dogs eat a lot of a lamb, sheep or deer, then the remains will smell of dog, but caution must be taken because carcasses that have been consumed by cats also smell musky, not too dissimilar to a dog. Footprints may be around the site. There may be no plucked belly hair under an opened belly. There generally will be more mess. Dogs do not have the need or time to systematically consume a whole animal carcass. So, most large animal carcasses that are found in places are more likely to have been consumed by a large cat or more.

Usually, other scavengers will come in at different times to take flesh away. Foxes are generally very suspicious of any deer carcass placed by humans or a carcass that a large cat has fed from. There is more about that subject in another chapter. Although we all assume dogs eat bones, in fact, wolves rarely eat the bones of large mammals and just pick off the flesh, yet a big cat will nearly always eat some of the larger bones. Cats need and eat bones more so than wolves or dogs especially female cats with developing young. They also eat far more skin, sometimes leaving a lot of flesh just simply to eat the skin! Ears, antlers and those parts are more attractive when certain minerals are needed.

Cats are also picky when eating and often leave bits and pieces in places with neatness. At times a cat eaten carcass will be scavenged by dogs whom are not bothered by the big cat smell and then, of course, two smells can be on the remains. Dogs will be attracted to any carcass they smell whilst out on a walk and try to home in on it if off the lead occasionally they will smell the big cat and show obvious alarm signs, or it may be curious or not care about a large cat smell and just plough into a carcass, taking off a leg, or the head maybe to take back to its owner.

When dogs and foxes interact with cat killed carcasses, they contaminate it with their own DNA.

Other Scavengers And Opportunists

I received a phone call to say that a fallow buck was found on somebody's front lawn in the New forest and asked if it were the work of a large cat. On examination, I concluded that indeed it was and took three samples of bone, nose and skin for analysis. There was no cat DNA, but there was fox, and the weirdest thing was the fact that the university technicians picked up contaminants such as otter and Burmese python! From the scissors used to cut the swab stick! So why no cat DNA? The buck was small and unhealthy, it also had a deformed antler and may have been lame, making it easy prey. It had been dragged under a holly tree and the rumen removed and placed behind the body.

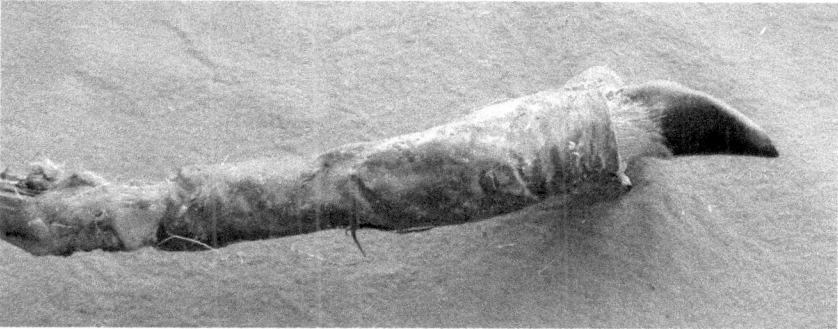

*Note that the hide had been pulled inside out, something that requires brute
strength and a typical large cat feeding sign.*

*The author with remains of three different deer carcasses. Large cats often have
special safe places where they drag their victim's body to consume in safety. A
build-up of remains can be found at certain eating places and the cat may drag or
carry the prey for quite a distance to reach one of its preferred feeding places. If a
cat is disturbed at a carcass, it may not return. This place had the remains of six
individual deer.*

Neil Roberts with a deer carcass eaten by leopards.

People also take deer and sheep. They can be killed by people for food, and when they do not, many other people suspect human intervention. In fact, it is possible that some people poach sheep or deer and blame it on dogs or large cats. Most poachers are big-time these days and do it wholesale, often to order. Whole herds of sika and fallow deer can be taken in a single operation using technical equipment and silenced rifles or crossbows, or other methods of killing quietly. Flyboy poachers take the main body but leave the heads and feet at a convenient dumping site such as a road lay-by on a quiet lane. Often piles of these can be found but soon get carried away by foxes. Some may be taken by large cats. As I have mentioned before and again, large cats will sniff out and scavenge carrion, especially deer carcasses. A few heads and feet left by poachers may well be eaten by large cats. Some dumping areas of dead animals may habitually attract all manner of scavengers for long periods of time and are great places to watch out for large cats.

Cats often carry food long distances to eat in peace or just a safe, quite favoured place, and in return, their own hard caught carcass remains may pile up and I have found several such places. This, of course, eliminates dogs as the killer usually or at least helps lean towards cats being the most likely candidate. Wolves do not consume a carcass in the same way as a cat, they rarely eat the ribs and when they do, they are messy and frayed and bent; they also do not eat as much skin or bones unless they are very small mammals. Another animal to bear into consideration is the wolverine (*Gluton gluton*). Similar in appearance to the Eurasian badger and a large martin, this animal is mainly a carnivore and is a very powerful animal, often predating on large deer. They are solitary mainly, like the American badger. There have been several reports of these animals in England in recent years. I have never found any evidence or seen any photos of evidence, but from witness reports, I have no doubt that some live around South Wales, Somerset and Devon.

Badgers

Badgers are misunderstood animals in many ways. The common badger, *Meles meles*, is a member of the Mustelidae family of weasel-like animals. Mustelids include weasels, stoats, mink, polecats, martens, otters and wolverines. They all have anal scent glands that are usually smelly and used for territorial markings. The polecat has the most notorious stinking fluid that erupts from pea-sized glands on either side of the tail. They are similar to a skunk in small and chemical makeup. The skunks are part of the same weasel family. Most of the weasel family are strict carnivores but with exceptions. The martins are partial to fruit and nuts, something that weasels and stoats never eat. Some polecats will eat some fruit rarely. Skunks also are more omnivores than carnivores. Badgers are one of several similar animals that have branched off from the weasel family to be strict omnivores or insectivores.

There are many badger type animals, but most of them live in small family groups or are basically solitary in existence for most of their lives. Eurasian badgers differ from their American counterpart in the way that they have become super social animals. American badgers (*Taxidia taxus*) are solitary and are slightly different in appearance but are more carnivorous in their diet. It is rare for carnivores to be social in their living habits. Many mongoose species are social but are mainly insectivores rather than carnivores and the types that have become large have also become solitary, such as civets. The reason why is because there just cannot be large social groups of carnivores; otherwise, they would decimate the areas where they live, killing and eating all other mammals and birds rather than insects and plants that cannot rejuvenate quickly to compensate.

Wolves, wild dogs and lions are the only true social carnivores on earth. Nature would not have it to have large carnivores living in large colonies. Badgers are omnivores. That means that they eat both animal and plant material, but even this is more refined, as they are more insectivores rather than the carnivores side. Most of any individual badger's diet consists of earthworms, plant bulbs, fruit and nuts. I have studied badgers since I was thirteen years old and have watched so many setts in different areas. I rarely saw any individuals eating other animals; it is an exception to the rule. I tried to feed badgers on all manner of meat, usually in the form of road-killed animals. Most of the time, they would simply ignore the carcass or move it out of the way if it was spoiling their sett area. There is always one badger within a colony that has a bit more of the weasel in it than its relatives and will dig up young rabbits and voles to eat. There is also another feeding behaviour trait of a less common form, and that is to eat chickens, ducks, or other small captive animals. That is where it usually ends regarding predatory instincts.

The more badgers within any area will throw up at least one individual of the rogue type animal, and then this is purely natural and good. In areas where worms are hard to get, because of drought, for example, then badgers will need to be more opportunists and may take to seasonably eating bird's eggs and chicks. Again this is usually an individual change and does not affect the whole colony. Boars are more likely to do so and satellite boars are even more so. There is always a boar that is destined to move around, spreading genetic diversity. Otherwise, badgers would be inbred. Badgers are strictly territorial and defend their areas by fighting and leaving territorial markers such as latrines along borders and scent marking. Badgers are usually habitual regarding where they urinate and defecate, and these areas are usually undercover, or near hedgerows or roads or other linear features.

Badgers are under great stress these days, as ever before but perhaps more so now that they have so many humans to contend with and our modes of transport. Roads and traffic are the badgers' main enemy (humans, really) and colonies are no longer few and far between, consisting of large colonies as they used to be, but many smaller colonies. This is because of how the huge death rate has shaped how they live. I once watched a colony of over sixty animals and another sett with at least that number and possibly more. It had two hundred and twenty entrances with forty or fifty in use at any one time. All these badgers ate worm's bluebell bulbs and corn cobs. I offered them dead crows, but they were offended. Squirrels were snorted at and fish would prevent them from coming out of their sett at dusk.

I have viewed thousands of video clips of animal carcasses placed in front of cameras. Never have I seen a badger consuming a deer carcass or any other animal placed there. I did once see an individual chasing rabbits, and not succeed! This does not mean that they will not eat a carcass if they are hungry or there is a drought and the ground may be too hard to dig, or worms have gone deeper. They will indeed if there is nothing else to eat, but usually, it is a last resort. In the very cold periods of snowy and icy weather, badgers may turn to scavenging, but again, it is an exception to the rule. I have looked at thousands of badger scats in latrines and rarely have I seen any animal bones, bird scales or dark blood scats. This does not mean that I haven't, just rarely and most of the time, never within most badgers colonies. I am sure that all badger experts would agree with me on this. Pheasant pens within a badger's territory may be hit by animals taking eggs and the odd young bird, but most of the time, badgers will not do any harm whatsoever around pheasant pens.

One thing that badgers do not do is kill sheep. One or two may take to eating newborn lambs, but this is rare and usually, the individual will be old and sick. This is not normal badger behaviour. And neither is scavenging. Some may tuck into a carcass if very hungry again in very cold or dry, or hot weather, but more often than not, badgers will eat the maggots of blowflies when they reach a large size.

Many people think that badgers are carnivores because they sort of know about them being bold and brave and used in fighting. One can also use a rabbit for the same reason, in fact, rabbits and hares spend their whole life fighting one another, but badgers don't. In fact, the humble badger is not even as hard as it is made out to be. Badgers have short jaws with teeth more like a pig. They have grinding molars for crushing nuts and bulbs, not bones and flesh. They have shortened canine teeth much shorter than foxes or even any other mustelid. This is for a reason, the reason being that they are not a carnivore, and nature did not intend them to be for all the reasons explained earlier. I think it is safe to assume the potential capabilities of badgers, but that is all, and many other things must be taken into consideration. So why do so many people include badgers when they debate carnivores killing deer or sheep. Even foxes cannot kill sheep and will not attempt it. Lambs are rarely taken by foxes unless they are a day old or are small. So badgers do not even enter the equation when assessing any carcass unless there have been exceptional circumstances such as a hard winter or hot, dry summer.

People must be educated regarding our wildlife. There is so much rubbish believed by uneducated people. Badgers do not spread bovine tuberculosis; the majority of people think they do. That is because they do not know badgers. Not many people do, and people are told lies by the authorities, as they always have done. But many people love to hate them and use them as scapegoats for people's own mess-ups!

Other Carcass Pickers

There are not many feral dogs within the UK. In areas of Europe, where they occur commonly, it will be even more difficult to separate dog kills from cat kills. There are reports of leopards or very big black cats in France, Belgium, Germany, Spain, Sweden, Switzerland, and Italy. The large cat phenomenon is also present in North America, Australia and New Zealand.

In North America, pumas are native and in the extreme south, the jaguar. There is no doubt that these animals are considered when it comes to analysing carcasses. In the North, though, large black cats may be melanistic pumas or leopards. The difference between America and the UK is the fact that there are two species of bears over much of the area whilst the UK doesn't have them, and some other smaller carnivores in the way of canids.

Coyotes and wolves, wolverines, two or three species of fox, American badgers and Fisher martin are the main other candidates for scavenging carcasses. In the UK, the main scavengers are (in order of most likely) carrion crow, magpie, raven, fox, domestic cat, domestic dog, Wild boar, Chinese Muntjack, polecat, American mink, pine martin, otter, blackbird, robin and all manner of small rodents.

Many or all of these animals suggested can alter the appearance of a carcass or add marks that could be wrongly analysed. A blackbird, for example, may decide to peck out plugs of flesh from a dead dear in such a way that it may look like teeth holes! And that is one of birds, right at the bottom of likely candidates so the other more likely candidates can do more damage to the carcass.

A muntjack, for example, may decide to eat some flesh and do so by hoisting it up, tugging out little pieces of flesh or skin. Generally, muntjack, like all other deer, are herbivores, but this family of deer are more primitive and has been known to eat dead animals sometimes regularly if available.

When buzzards feed, they often cause the skin to fall back, making it look like a cat had sheared off some skin with its carnassial teeth. Ravens can polish off a dear carcass quite efficiently but will not eat any bones. The crow and magpie also will not eat bones; in fact, most birds do not eat bones as it could spell disaster. Birds only have one exit hole at the anus and faeces and urine are ejected through it. Both need to be combined. Any rough indigestible material must be ejected back through the way it came in, through the mouth. Bird pellets are simply these large amounts of fur, feather and bone that cannot pass the other end. Birds are basically reptiles with adaptations in some ways but retention of other body parts and functions.

Most birds of prey and gulls, herons, and corvids must reject waste by forming pellets. These, in turn, can easily be confused for droppings and even large cat scats. The bird pellet is basically circular in shape or slightly long. It is formed within a bird's simple stomach and compacted. A scat formed within the stomach of a carnivore is long and often twisted. In a bird pellet, the items are usually gathered together in tight compaction so that the whole skeleton of, say, a mouse will be visible within all the fur from that single individual prey item or others eaten within the same day or night.

I often find remains of large water birds that have been predated on. Swans and Canada geese, along with herons, feature regularly in the diet of some of the large cats within my study areas. I suspect all three large cats to be responsible. Around the Poole harbour are many birds and during the winter, many migrant birds such as Brent geese, greylag geese, shell duck, mallard, gadwall and other duck species fill the water. This attracts Lynx and puma. The leopard could swim out to capture swans on little islands. They seem to have no problem with killing deer that are out in small islands with much water in between and marshland. Here just a few of the largest feathers are all that remain of this swan. The primary feathers have been bitten at the base, consistent with a carnivore as opposed to a raptor. There have been reports from other areas of England of swans being stashed in trees where other signs and sightings of leopards have been made. Swans are often on the menu for leopards and they may swim out to get them as they roost on little islets of trees within the large lakes. Swans are large, heavy birds, but the cats retrieve them and bring them back some distance away from the water to consume under bracken. Not much is left.

If buzzards, eagles or even herons have eaten from a carcass, they may throw up a pellet by the carcass. Buzzard and ravens can eat bones the size of rabbit legs but nothing larger. Eagles can eat larger bones and they will show in pellets. Basically, a bird will not eat bones that cannot fit into its mouth and they cannot break bones with their beak except the breast bones of many birds. A bird's beak is never strong enough to break bones in the same way that carnivorous mammal teeth can. A cat will eat most bones if it can, and in small mammals like squirrels, rabbits, fox, cubs, etc., all the bones will be consumed usually, but there is some individuality in which each cat consumes its prey. A lynx, for example, may not like eating rabbit pelvises and bite them out, leaving them along with the furry feet. Another lynx may like frogs but leave the tiny reproductive organs. A domestic cat may like most of the mouse except the head or heart.

Also, a large cat may sheer the thin membrane of muscle which adheres to the skin. As it eats the main body, it pulls the skin back either by tugging it off or simple by sheering it off. When the latter happens, one can see striations on the inner side of the skin, which can look as though a person has skinned it with a blunt knife.

When a cat eats, it is at great risk of losing many of its whiskers, as they are usually long, especially in the case of leopards, which usually have longer whiskers than any other cat. This is because of their nocturnal hunting, tree climbing capabilities and the need to judge distances. The whiskers can be caught between the carcass and the teeth, and so many will be cut through and be within the carcass or eaten by the cat. The latter is more common, and a search through scats, often throw up not just whiskers of the prey but of the cat itself.

In scats, generally, one can find the hairs of the original owner because they are fastidious cleaners and often ingest their own fur, especially when moulting. Because the cat's whiskers are so long, it puts them at great risk of losing a few on a regular basis. They fall out and grow back quickly on a regular basis. The cat may pinch them within the carcass, such as the ribs, or if it is eating the head. The whiskers often stick to the flesh and can be hard to spot; often, a dry carcass will reveal them easier. I have also found cat whiskers on barbed wire fences. A cat can feel its whiskers, of course and usually knows if they are trapped, but accidents often occur and at the rate that they grow back, do not cause sufficient inconvenience. The whiskers of a cat are sensory organs used to detect minute changes within the electromagnetic field. They can pick up electrical devices and the electricity generated by other animals, including temperature. Cats are more likely than any other mammal to have white coloured whiskers. Half-light and half dark are also very common and most cats will have both forms.

Most other British mammals have mainly dark whiskers, but some have bi toned whiskers, especially hare, rabbits, mustelids and the pseudo whiskers of deer. Deer have extrasensory hairs that are distributed over the body, on the sides of the belly, neck, top of the front legs. These are intermediate between true whiskers and normal hairs. They are longer than normal hairs and serve to be used like sensitive whiskers. Deer have long whiskers around their eye mainly and just a few short ones around the muzzle and longer ones on the chin. They are not made up of the same structure as true whiskers on the muzzles of most wild animals.

Most of the hairs found on and around a carcass will be from the dead animal itself, but there may be moulting hair from the predator or scavenger. In the case of deer, all deer hair is very similar and general knowledge of the typical shape and colour of deer hair comes in handy, not just here but at all times. Most hair surrounding the dead body will be of a similar appearance and the bases of all the body hair of deer look fluffy and pale grey or white. Deer do not have as much underfur as all cats do as well as most other mammals. Only on the head can underfur be found on deer species, especially fallow, sika and red. The underfur is dense and short, usually with longer outer or guard hairs of a more sparse distribution. The long guard hairs of a cat stand out among the typical grey-based crimped hairs of a deer. Cat hairs are usually thin and shiny. In the case of most large cats being black, then the colour alone may stick out on the carcass or surrounds or even the bones.

A broken spine is often evidence of a strong animal and may even suggest killing methods. A broken rib or two by the heart may also suggest an attacking blow. Deer may be hit and killed on a nearby road and then bones may be shattered. Usually, limb bones are broken once, but the break is different to the way in which a large cat would break it. The cat sheers it off and the fracture will rarely be a long one but a short bite consisting of a cleaner break with trauma to the edges. There will not be the domino shattering (breakage due to shockwaves).

Another adult fox carcass within an area with many sightings by the public. In this field, many deer carcasses are found. It was a shame that the snow fell after the cat left the scene. Footprints were not found other than pheasant and raven.

Usually, a large cat will kill its prey and then drag or carry it away for safe consumption, often at a favoured site out of harm's way. This is not always the case, as many cats will kill and eat a sheep at the same spot. I think it depends on the situation at the time and whether or not the cat feels safe. It also depends on the stock and whether or not they are spooked or relaxed. It is not in the cat's best interests to get its prey afraid of it! Dogs do not drag carcasses but can move them around accidentally by tugging at the flesh. Foxes do the same and so do badgers on the very rare occasions when they consume carrion.

Foxes can and often do carry away small lamb carcasses when feeding cubs and will carry geese and ducks, pheasants etc. Usually, they will lessen the load by plucking the larger feathers out by yanking them out with incisor teeth, leaving typical puncture and pulling trauma, or biting off the wings when consuming it and leaving typical linear shearing in a similar way to how a cat does it.

The typical eaten out remains of a lamb from Farmer John Garthwaite from Devon. The cat made two nightly visits to finish the ewe.

Pumas often mark their kill sites with scats, but usually, the carcass will be buried when it is not eaten. I have not found scats by suspected leopard kills except those of cubs actually at the carcass.

Foxes often try to take claim to a carcass by leaving scats. I have found some puma-like scats at kill sites. Obviously, carcasses in trees are usually the doings of leopards; it remains may stay in place for years before falling to the ground. They will normally have the telltale tooth marks of voles, mice and squirrels on some bones.

Cub cats play with carcasses and they often get torn around, broken up, hidden up trees, put into holes, under woodpiles. A person may even think that a person or people had been responsible for messing around. I once was shown photos of pieces of deer pelt lodged in tree branches. I suggested that a few cubs had been whizzing around with the larger remains of the pelt, but it has snagged several times, ripping off small pieces. People did not believe me! You only have to watch some of the high-quality wildlife documentaries on television to learn things like that.

Antler Eating

Large cats eat deer antlers. Antler is only bone but of a difference or two. It is a fast-growing bone that starts out as soft, flexible cartilage before slowly hardening as the antler grows from the base upwards. Deer antlers are for fighting and display, so they need to be as large as possible and rightly formed, equal if possible, yet the individual deer has no control over it, only by its hormones that can cut on or off the supply of nutrients and hormones.

As antlers grow, they are full of nutrients taken in by the plants and minerals from the earth. They may take several months to grow before hardening and then being used for fighting before dropping off later.

The large deer species in the UK, i.e., red fallow and sika, all shed in March and April, older bucks shed in February. The antlers start to grow immediately covered by a skin of velvet, fur basically very short and velvety. The summer growth allows swift development, and by

September, they are full-grown and the velvet falls off as they harden, ready for the autumn rut.

Roe bucks shed in November and re-grow during the winter months; hence they are much smaller and are covered by much longer fur. When the antlers are growing, especially soon after they start growing, they are like a delicacy for carnivores and like a multivitamin box. A lactating female cat will naturally home in on this essential source of calcium and other minerals to help her cub's development.

I have found that on most stag or buck deer carcasses with antlers, there had been at least tooth trauma on the antlers and in the case of soft growing antlers, eaten down to the stumps or one antler consumed and the other partially. I have even found hard antler partially eaten. Antler is extremely hard, more so than normal bone, so it takes much tooth and jaw power to break them. Foxes may eat young growing antlers or naw on harder ones if no flesh is available. Deer also eat antlers, but in a slow way, they pick them up and carry them around, scraping small amounts before dropping them, but over time where hundreds of individual deer or the same few picking at it grinds it down to nothing over time.

Rodents also eat bones and antlers and, over a long time period, scrape away the calcium with typical striations until holes appear and the bones or antlers look very misshapen until they are consumed. Cats also eat ears and many a time, skin and ears seem to be the only parts consumed. It may be, that a lactating female feeds on the most important parts before being disturbed. Foxes eat ears also but usually as a last resort, concentrating on proper flesh first. So a carcass with obvious clean ear or antler removal is most likely to be the result of large cat feeding, especially if all the other hallmarks are obvious. Some of my friends and colleagues have handed me shed antlers with large cat tooth trauma on them! And I have found a few myself, this also means that cats pick up antlers and may bite them or try to eat them. I do not find enough carcasses of muntjack to know the situation regarding them. This species wasn't common in Dorset until quite recently and even then, I rarely came across remains of cat eaten carcasses even in areas where they occurred in my study areas! This may be because of its small size and the ability of a leopard to consume all bones? Many of my colleagues, especially in the Midlands and Gloucester, regularly find muntjack remains, this may be because it is the most common deer species in this area, second to roe and then the larger species. In Dorset and the New forest, it is the larger deer species that I usually find as being large cat victims.

This roe buck had its antlers wedged in a tree and was eaten out by a possible leopard. Stroud district Gloucestershire.

The Growling corner leopard was responsible for this stag being consumed, he ate the rump, but later at least one other cat came to fully polish off the rest of the carcass.

Cats On Cats

Cats eat cats; this is a law of nature. Big cats will usually kill and eat smaller cats as it eliminates competition. Even if the small cat is not a direct threat to a larger species' food supplies, to cats, all cats are the same in some respects. A puma will often catch and kill a lynx or bobcat. The bobcat or lynx will catch and kill a domestic moggie. A leopard will catch, kill and maybe eat any other cat smaller than itself! In some recent puma studies, biologists confirmed that male pumas will habitually kill and often eat other cubs and even females, especially young male pumas. This is how cats regulate their numbers. There will be bodies of cats found with the same hallmarks of an eaten out deer. The larger cat may leave a victim dead or take some flesh from it before discarding it, or it may be hungry enough to consume it in the same way as it would a deer carcass. There will be no rumen to remove, so the internal parts will be eaten, or a large area of skin and rump, for example. Often a whole litter of cubs will be killed and left around the den site; One or two may be carried off and eaten.

Updated 2020.

I have had what could be evidence of a puma or lynx being eaten by a larger cat. Somewhere in the UK, a body was found in a field that had all the hallmarks of being eaten by a large cat. It was turned inside out, flesh consumed and some bones. The body was left in the sun to bleach and dry out.

Safe feeding places (SFP)

I have found areas where multiple skeletons pile up! One such site and my first, was on Studland heath, quite near a used footpath through the heath and mire. It was within a strip of willow Carr bog adjoining thick heath. The area was between wet, humid bog and dry, warm sandy heath. It wasn't a particularly good stalking area, but nearby were many a good places where deer congregate to drink, high bracken, gorse, heather and purple moor grass, the best hunting habits on the heathland sites.

The area where the bones were, was devoid of vegetation but bare ground. I found three skulls of sika hinds, all about the same size and the spines and broken down ribs. There were several leg bones and vertebrae scattered. All animals would have been eaten within a time scale of about six months. The bones were clean but not growing algae or moss. The bones were in semi-shade, so some had started to sun bleach; others were not. They had been consumed during the summertime, when ants and maggots clean the bones and decomposition is short. There were some tooth pits on some of the bones, especially the pelvic girdle and they matched with the dentition of a leopard. I collected some of the bones and hid the rest so as not to attract people or dogs.

Later in the following winter, I found two more carcass remains, again similarly sized sika deer, yet one was a yearling stag. I then realised that this must be a safe feeding place (SFP). I have found no less than seven similar places where multiple skeletons build up. I have found many old sites with bones that are more than a couple of years old and up to ten years old but with no recent bones, and others where all bones are recent, and some areas where there may be spates of usage over a couple of years, but perhaps short time periods of a few months within each year.

Habitually cats will find these safe areas where they know that they can consume a carcass without being disturbed. The cat needs as much of the body as possible and with so many disturbances of cat feeding, they will use and re-use these quiet sites to enable them to get the most out of the carcass. Often they are in similar areas, perhaps an old bomb crater of a small coomb, an isolated stand of thick gorse or behind someone's garden shed! Where leopards are concerned, large trees such as oak or pine may be used to store prey, but I don't think that they consume them in trees; but prefer to bring them down to feed, often stashing them back up to finish later or to stop foxes from eating it, whatever, they are often left up there to break down and then the bones fall when windy or are scattered by squirrels. Still the remains will pile up under the trees. Other researchers have also found SFPs often in thick conifer woods, quarry cliffs, reed beds, railway lines, water treatment works, electricity stations, or out of bounds areas for people or dogs.

During the quiet night, a cat may feel secure enough to feed where it kills its prey, even if in the most unlikely of places such as roadsides or people's gardens. I once found the remains of an otter on someone's front lawn bordering a busy main road; although the otter may have initially been a roadkill, it had the typical hallmarks of being consumed by a large cat. Twice I have been notified of deer remains in gardens, and this has puzzled and alarmed the human occupants. Many people actually see large cats killing animals in their gardens or farmland close to buildings. Sometimes the people hear a racket at night only to find the remains in the morning. At other times the people find the remains of their pets! Such as cats, dogs, rabbits or goats. Often people see cats taking fish from ponds or find the remains and sometimes assume an otter was the culprit.

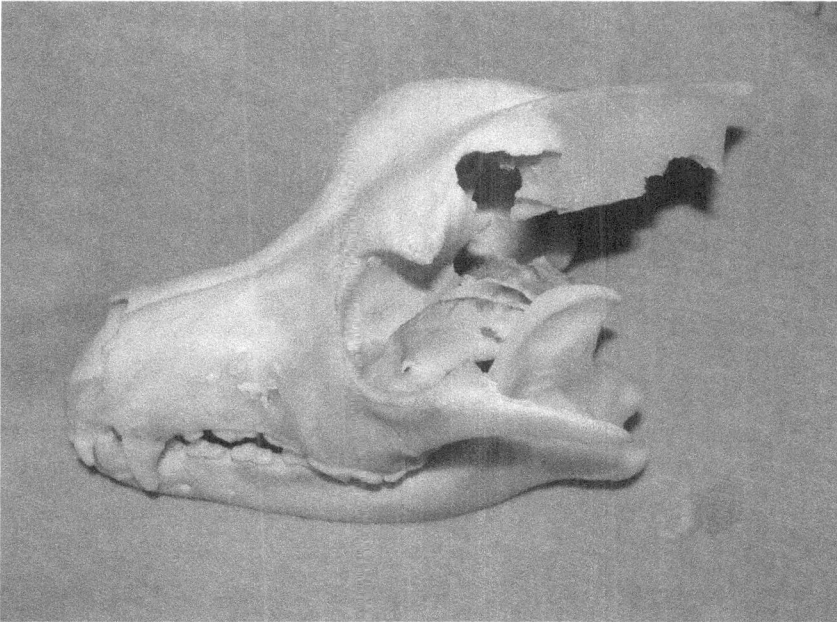

Dogs are often taken throughout the UK, just as one would expect from the other natural countries of origin for leopards, less so for pumas. This skull was found in a typical area where much other large cat field evidence persists. The cranium has been removed and a very large canine tooth hole is clearly visible. This collie did not return home after wandering off as it often did. It was found by a delegate at the Dorset big Cat conference in 2014.

Often a large cat will carry its prey into the middle of a field to consume and feel safer there than anywhere else. Often at night, when quiet, a cat will do this and often drag road-killed animals into a field to eat; usually, it may be right in the middle of the field as the cat will have a full all-round view of any danger. The cat can then flee in any direction away from the source of concern.

Recently I have heard of many reports of SFA in all parts of the UK. Although many researchers may not know what they are, some think that people throw all the bones in certain areas to rid them, or others think that it is the exact place where the large cat had killed them, but in reality, the cat may have carried them for over a mile! From the original kill place.

Tunnels and mines and caves are good places to find these bone pileups, and I have seen photos of dozens of sheep remains or deer in old army sheds or Nissan huts. I have always assumed them to be mainly from Pumas rather than leopards, especially in my areas, as I find that leopards prefer to stash them in trees. I have found more puma footprints around the areas of SFP rather than leopards, but this is not always the case as some of them are definitely attributed to leopards, especially on heathland.

It is always large animals that are taken to these sites as I have never found the bones of smaller mammals or birds at them and they usually consist of deer remains. Others found SFP containing just sheep or a mixture of deer and sheep. Occasionally one will hear about several badger carcasses at such sites. Some people may have got the wrong idea about finding the remains of badgers and foxes at these sites and assume that a gamekeeper or farmer had shot the animals before dumping them at a convenient tip! And indeed, this does happen, but to muddy the water, I came across a site on the Dorset Hants border where a sheep farmer regularly threw his dead ewes and sometimes rams over the fence into a wood on top of a badger sett. Maybe he thought that the badgers ate then and cleaned them up! Well, the badgers never touched the dead sheep except dry to drag them away from the entrance holes. In fact, a puma often came to feed. At times one or two whole late lamb or ewe carcasses would be placed behind a fence; usually, they had been killed by dogs, an alarming commonplace atrocity to happen all too often these days. After three days, I would return to find both carcasses stripped out in large cat fashion. There would be tooth pitting typical of puma and often scats left nearby. I wonder if the farmer actually knew about the feeding puma or really thought that the badgers ate them. Well, to be honest, it is most likely the latter!

The remains of a badger skeleton lay just a few meters from a suspected deer carcass. There were no other bones from the skeleton around and so I assume that they were consumed. There are teeth marks on some of the bones and others have parts cleanly bitten off. (Not shattered by road vehicles). The ribs have also been bitten down to the base at the front of the thorax and just halfway down on the hind part of the thorax. This area is within the territory of at least one large black cat that was seen by many people around the Northern part of Christchurch and Bournemouth airport.

AFTERWORD

Go to hangaripublishing.com to learn more about the Author and stay up to date with their newest releases.

www.ingramcontent.com/pod-product-compliance
Lightning Source LLC
Chambersburg PA
CBHW070110030426
42335CB00016B/2094